For more than a quarter century, the annual *Esri Map Book* has showcased a variety of outstanding maps our users have produced to help enhance their success. This year is no different. Nearly 100 organizations and entities are featured in the 28th volume, and almost a dozen languages are represented in maps covering virtually all regions of the globe.

And again, as in past years, the promise of geographic information systems (GIS) and geography as an integrated platform is aptly demonstrated. The science of spatial thinking and reasoning is pervasive as cloud, web, server, desktop, and mobile technologies and applications take the power of GIS to the farthest reaches of our world.

In the end, though, it's the *people* working with the software who make all the difference. Their vision, creativity, education, and knowledge—coupled with ever-increasing strides in GIScience—are the real keys behind what you see in these many fine maps.

I am grateful to the GIS professionals and students who have worked so conscientiously to produce the images featured here; they are proof that innovative cartography, thoughtful spatial analysis, and the inspired visualization of geographic data are fundamentally important to making the planet a better place to live.

Warm regards,

Jack Dangermond

A Letter from
Jack Dangermond

Table of Contents

Spatial Analytics for Marketing

Mosaic
Irving, Texas, USA
By David Drury

Contact
David Drury
david.drury@mosaic.com

Software
ArcGIS 10 for Desktop

Data Sources
Mosaic, Esri, DeLorme, NAVTEQ, USGS, Intermap, iPC, NRCAN, Esri Japan Corporation, METI, Esri China (Hong Kong Limited), Esri (Thailand) Co. Ltd.; TomTom, 2012

With limited marketing budgets and increased competition, it is critical that retailers carefully target their marketing expenditures to ensure they are achieving the best return on investment. This series of maps provides an overview of Mosaic's spatially driven analysis methodology that uses ArcGIS to lead retailers to the best locations to offer products, promotions, and store events. Mosaic is a marketing agency that connects clients with consumers at retail and online levels and works with many Fortune 100 companies.

In this analysis, the Mosaic GIS Team determined the best store locations to launch a new consumer electronics product. ArcGIS solutions, including Esri Business Analyst Desktop, provided the platform to integrate historical sales data, in-store traffic volume, competitor locations, and consumer demographic profiles into an analytic model that identifies store locations with the best return on investment for the product launch.

Courtesy of Mosaic.

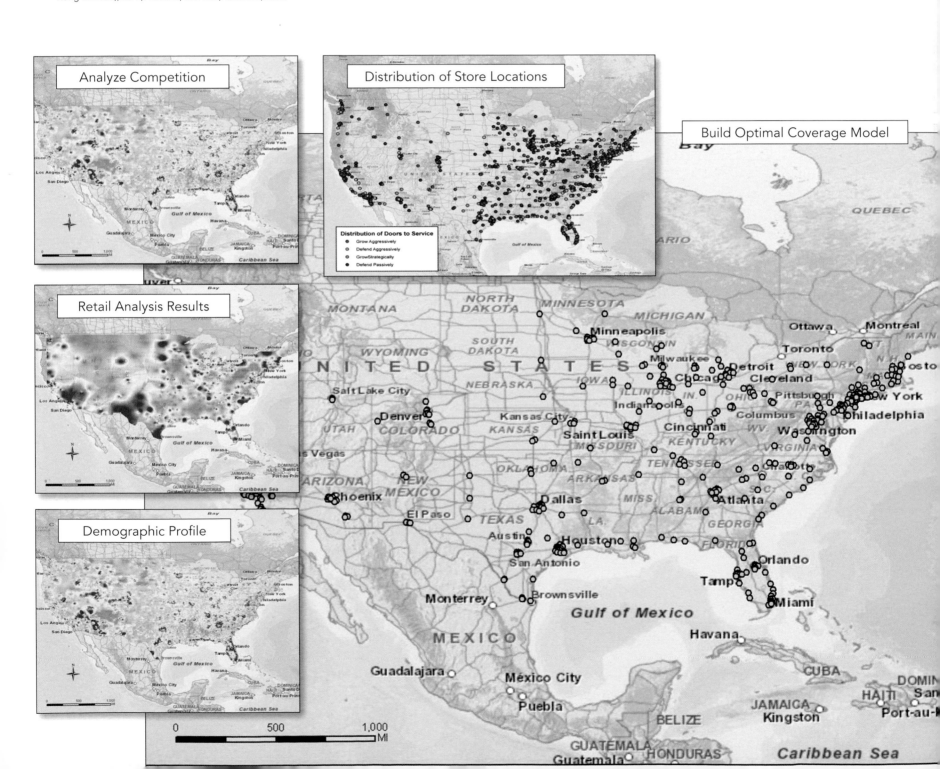

Cartographic Excellence in Creating a Community Basemap

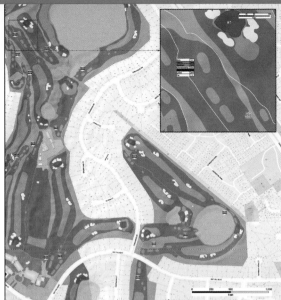

Orange County Property Appraiser
Orlando, Florida, USA
By Manish Bhatt, Frank Yang, and Dan Duckworth

Contact
Manish Bhatt
mbhatt@ocpafl.org

Software
ArcGIS for Desktop, ArcSDE, ArcGIS for Server
(Silverlight API)

Data Source
Orange County Property Appraiser

The Orange County Property Appraiser is responsible for identifying, locating, and fairly valuing all property, both real and personal, within the county for tax purposes. The objective of these maps was to enhance the cadastral map with value-added cartographic features, transforming the traditional tax map from a flat product to a rich cartographic community basemap. These multilevel cached images construct a common basemap for all GIS applications, viewed by 5.2 million web users in 2012.

Courtesy of Orange County Property Appraiser, Florida.

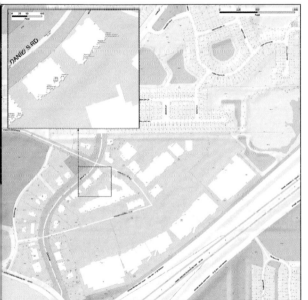

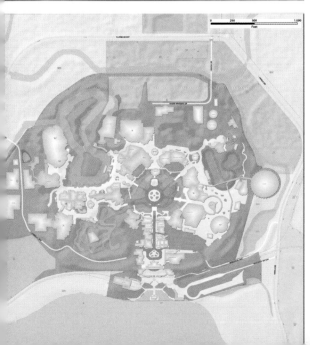

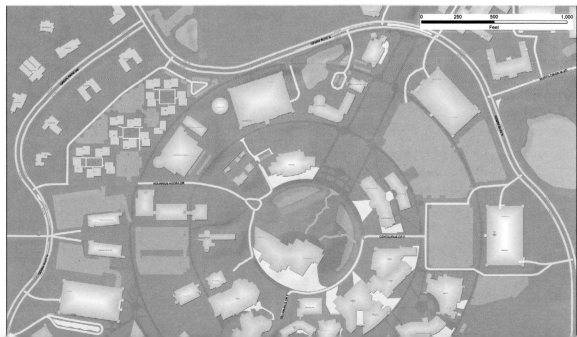

Regional Potentiometric-Surface Map of the Great Basin Carbonate and Alluvial Aquifer System in Snake Valley and Surrounding Areas

US Geological Survey, Nevada and Utah Water Science Centers

Carson City, Nevada and Salt Lake City, Utah, USA

By Philip M. Gardner, Melissa D. Masbruch, Russell W. Plume, Susan G. Buto, and Joseph F. Gardner

Contact
Susan Buto
sbuto@usgs.gov

Software
ArcGIS 10 for Desktop, Adobe Illustrator

Data Source
USGS digital data

Water-level measurements from 190 wells were used to develop a potentiometric-surface map of the east-central portion of the regional Great Basin carbonate and alluvial aquifer system in and around Snake Valley, eastern Nevada, and western Utah. The map area covers approximately 9,000 square miles in Juab, Millard, and Beaver Counties in Utah and White Pine and Lincoln Counties in Nevada.

Recent (2007–2010) drilling by the Utah Geological Survey and US Geological Survey (USGS) has provided new data for areas where water-level measurements were previously unavailable. New water-level data was used to refine mapping of the pathways of intrabasin and interbasin groundwater flow. At twenty of these locations, nested observation wells provide vertical hydraulic gradient data and information related to the degree of connection between basin-fill aquifers and consolidated-rock aquifers. Multiple-year water-level hydrographs are also presented for thirty-two wells to illustrate the aquifer system's response to interannual climate variations and well withdrawals.

Courtesy of US Geological Survey.

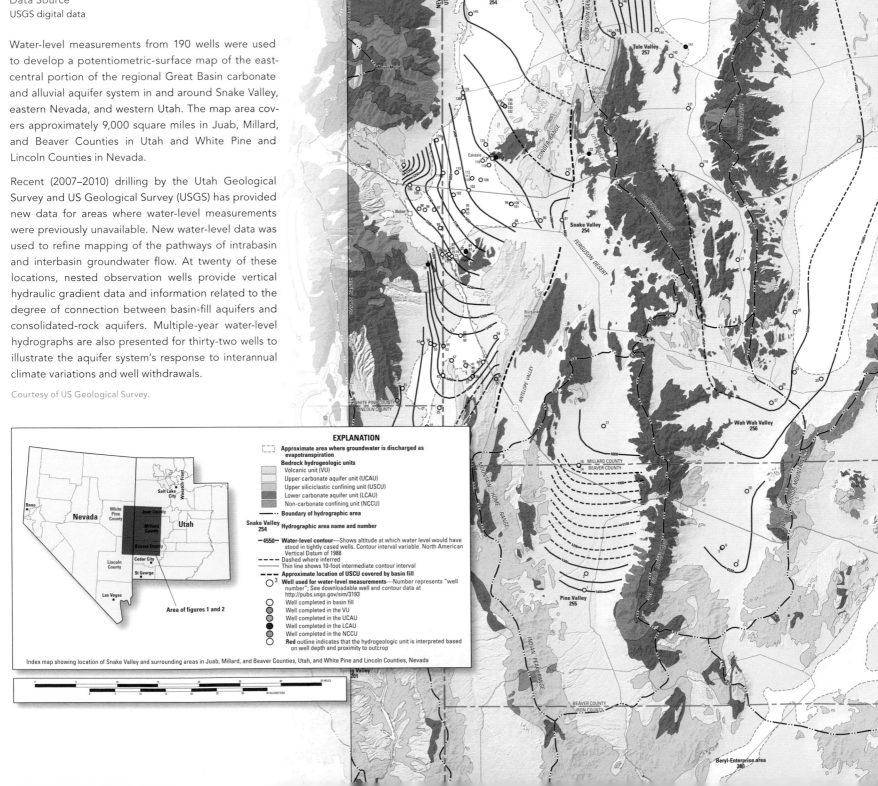

EXPLANATION

Approximate area where groundwater is discharged as evapotranspiration

Bedrock hydrogeologic units
Volcanic unit (VU)
Upper carbonate aquifer unit (UCAU)
Upper siliciclastic confining unit (USCU)
Lower carbonate aquifer unit (LCAU)
Non-carbonate confining unit (NCCU)

Boundary of hydrographic area

Snake Valley
254   Hydrographic area name and number

—4550— Water-level contour—Shows altitude at which water level would have stood in tightly cased wells. Contour interval variable. North American Vertical Datum of 1988
Dashed where inferred
Thin line shows 10-foot intermediate contour interval

Approximate location of USCU covered by basin fill

○³ Well used for water-level measurements—Number represents "well number"; See downloadable well and contour data at http://pubs.usgs.gov/sim/3193

○ Well completed in basin fill
○ Well completed in the VU
◑ Well completed in the UCAU
● Well completed in the LCAU
◓ Well completed in the NCCU
○ Red outline indicates that the hydrogeologic unit is interpreted based on well depth and proximity to outcrop

Index map showing location of Snake Valley and surrounding areas in Juab, Millard, and Beaver Counties, Utah, and White Pine and Lincoln Counties, Nevada

Area of figures 1 and 2

Northeast Negev (Ha-Makhteshim) Region Topographic Map

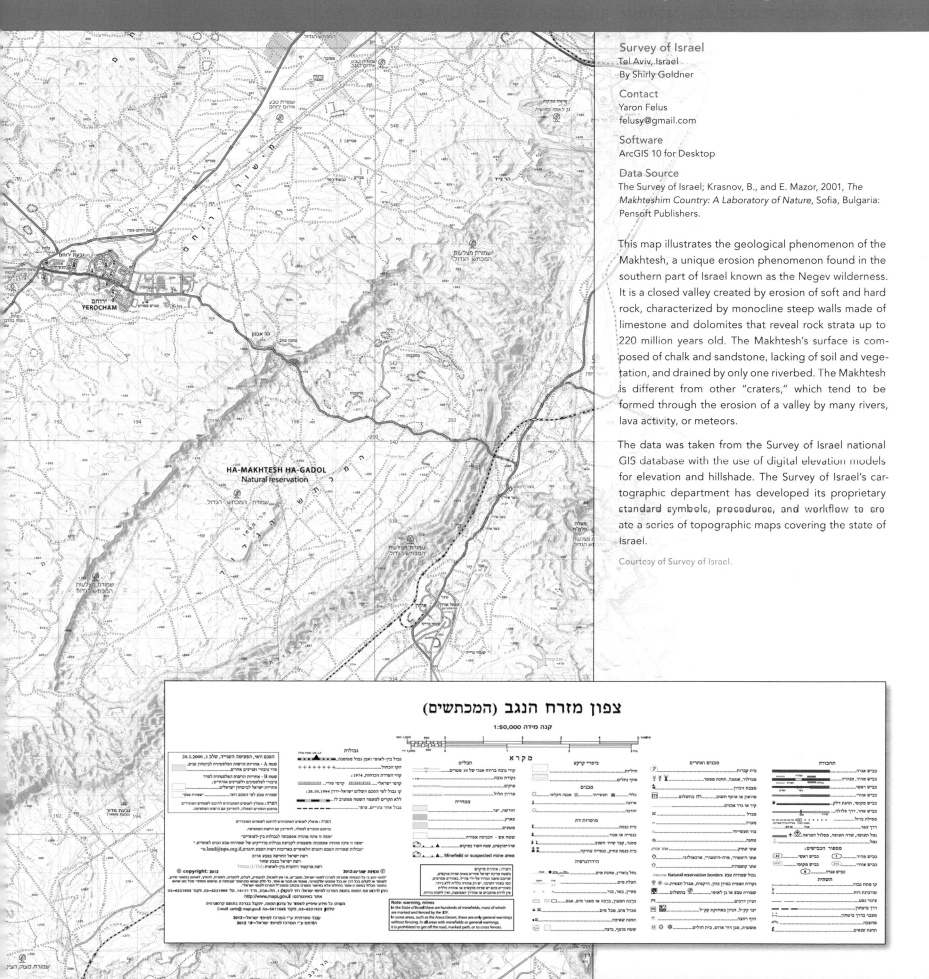

Survey of Israel
Tel Aviv, Israel
By Shirly Goldner

Contact
Yaron Felus
felusy@gmail.com

Software
ArcGIS 10 for Desktop

Data Source
The Survey of Israel; Krasnov, B., and E. Mazor, 2001, *The Makhteshim Country: A Laboratory of Nature*, Sofia, Bulgaria: Pensoft Publishers.

This map illustrates the geological phenomenon of the Makhtesh, a unique erosion phenomenon found in the southern part of Israel known as the Negev wilderness. It is a closed valley created by erosion of soft and hard rock, characterized by monocline steep walls made of limestone and dolomites that reveal rock strata up to 220 million years old. The Makhtesh's surface is composed of chalk and sandstone, lacking of soil and vegetation, and drained by only one riverbed. The Makhtesh is different from other "craters," which tend to be formed through the erosion of a valley by many rivers, lava activity, or meteors.

The data was taken from the Survey of Israel national GIS database with the use of digital elevation models for elevation and hillshade. The Survey of Israel's cartographic department has developed its proprietary standard symbols, procedures, and workflow to create a series of topographic maps covering the state of Israel.

Courtesy of Survey of Israel.

Cenomanian Global Palaeogeography

Getech Group plc
Leeds, United Kingdom
By Kevin Mckenna, Paul Markwick, and Peter Birch

Contact
Kevin Mckenna
kmk@getech.com

Software
ArcGIS Desktop 9.3.1

Data Source
Getech Group plc

Getech's Atlases of Global Palaeogeography provide a digital representation of the earth's surface through time, including the reconstruction of past depositional environments, sediments, elevation, rivers, and bathymetry. Any digital spatial data can then be reconstructed onto these maps for use in a wide range of applications, from oil, gas, and minerals exploration to paleobiogeography, to plate tectonics and geodynamics, to the very latest paleo-climate, -ocean, -tide, -ice, and -vegetation modeling. This is the first time that every stratigraphic stage in the geological record has been mapped in this detail.

These maps have been built entirely in ArcGIS and are underpinned by extensive databases of gravity and magnetic data, structural and tectonics, crustal types, and geology. ArcGIS not only allows all these elements to be displayed and queried spatially at all scales but also facilitates updates and improvements to be built relatively quickly and passed to users.

These atlases form a core part of Globe, Getech's innovative New Ventures Exploration toolset, which is currently being used by over ten of the world's largest oil and gas exploration companies.

Courtesy of Getech Group plc.

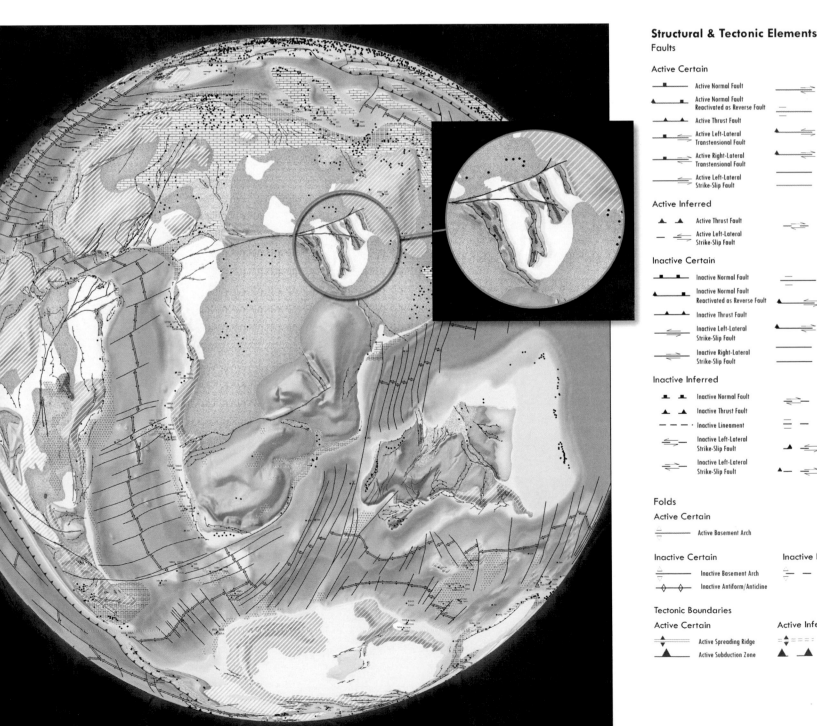

Structural & Tectonic Elements

Faults

Active Certain

- Active Normal Fault
- Active Normal Fault Reactivated as Reverse Fault
- Active Thrust Fault
- Active Left-Lateral Transtensional Fault
- Active Right-Lateral Transtensional Fault
- Active Left-Lateral Strike-Slip Fault
- Active Right-Lateral Strike-Slip Fault
- Active Undifferentiated Strike-slip Fault
- Active Left-Lateral Transpressional Fault
- Active Right-Lateral Transpressional Fault
- Active Undifferentiated Fault
- Active Lineament

Active Inferred

- Active Thrust Fault
- Active Left-Lateral Strike-Slip Fault
- Active Right-Lateral Strike-Slip Fault

Inactive Certain

- Inactive Normal Fault
- Inactive Normal Fault Reactivated as Reverse Fault
- Inactive Thrust Fault
- Inactive Left-Lateral Strike-Slip Fault
- Inactive Right-Lateral Strike-Slip Fault
- Inactive Undifferentiated Strike-Slip Fault
- Inactive Left-Lateral Transpressional Fault
- Inactive Right-Lateral Transpressional Fault
- Inactive Undifferentiated Fault
- Inactive Lineament

Inactive Inferred

- Inactive Normal Fault
- Inactive Thrust Fault
- Inactive Lineament
- Inactive Left-Lateral Strike-Slip Fault
- Inactive Left-Lateral Strike-Slip Fault
- Inactive Right-Lateral Strike-Slip Fault
- Inactive Undifferentiated Strike-Slip Fault
- Inactive Left-Lateral Transpressional Fault
- Inactive Right-Lateral Transpressional Fault

Folds

Active Certain

- Active Basement Arch

Inactive Certain

- Inactive Basement Arch
- Inactive Antiform/Anticline

Inactive Inferred

- Inactive Basement Arch

Tectonic Boundaries

Active Certain

- Active Spreading Ridge
- Active Subduction Zone

Active Inferred

- Active Spreading Ridge
- Active Subduction Zone

Wastewater Treatment Process OCSD Reclamation Plant No. 1

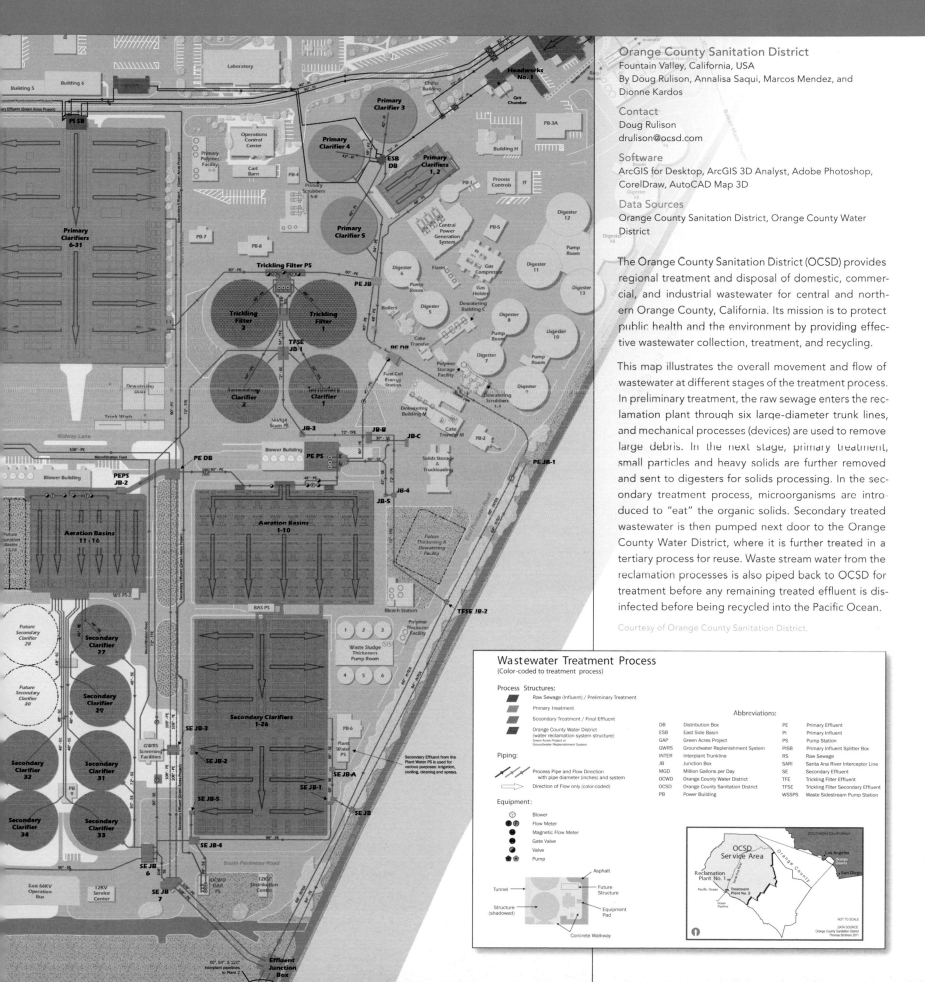

Orange County Sanitation District
Fountain Valley, California, USA
By Doug Rulison, Annalisa Saqui, Marcos Mendez, and Dionne Kardos

Contact
Doug Rulison
drulison@ocsd.com

Software
ArcGIS for Desktop, ArcGIS 3D Analyst, Adobe Photoshop, CorelDraw, AutoCAD Map 3D

Data Sources
Orange County Sanitation District, Orange County Water District

The Orange County Sanitation District (OCSD) provides regional treatment and disposal of domestic, commercial, and industrial wastewater for central and northern Orange County, California. Its mission is to protect public health and the environment by providing effective wastewater collection, treatment, and recycling.

This map illustrates the overall movement and flow of wastewater at different stages of the treatment process. In preliminary treatment, the raw sewage enters the reclamation plant through six large-diameter trunk lines, and mechanical processes (devices) are used to remove large debris. In the next stage, primary treatment, small particles and heavy solids are further removed and sent to digesters for solids processing. In the secondary treatment process, microorganisms are introduced to "eat" the organic solids. Secondary treated wastewater is then pumped next door to the Orange County Water District, where it is further treated in a tertiary process for reuse. Waste stream water from the reclamation processes is also piped back to OCSD for treatment before any remaining treated effluent is disinfected before being recycled into the Pacific Ocean.

Courtesy of Orange County Sanitation District.

Missoula Floods—Inundation Extent and Primary Flood Features in the Portland Metropolitan Area

Oregon Department of Geology and Mineral Industries
Portland, Oregon, USA
By William J. Burns and Daniel E. Coe

Contact
Daniel Coe
dan.coe@dogami.state.or.us

Software
ArcGIS 10 for Desktop, Adobe Photoshop CS5.1,
Adobe Illustrator CS5.1

Data Sources
DOGAMI, US Army Corps of Engineers; Clark County (Washington), Washington Department of Natural Resources, National Agriculture Imagery Program, Oregon Department of Transportation

Between 15,000 and 18,000 years ago, continental glaciers formed a dam that blocked the Clark Fork River in Montana. This dam resulted in the formation of Glacial Lake Missoula, which contained 530 cubic miles of water. When Lake Missoula waters breached the ice dam, some of the largest floods known discharged nearly 350 million cubic feet per second—over 1,000 times the average discharge of the current Columbia River. This dam-and-breach process was repeated at least forty times over 3,000 years as the ice sheet advanced and retreated. With each breach, huge volumes of water raced across eastern Washington, eroding and depositing material before converging into the Columbia River Gorge. Floodwater rushed out of the constricted Columbia River Gorge and entered the Portland Basin, creating much of the large-scale geomorphology that exists today. Features include huge primary channels that are, in some places, scoured down to bedrock, large sand and gravel bar deposits, and extensive fluvial sculpting throughout the basin.

Lidar data in this map has been colored and shaded to reveal relative and absolute changes in elevation within the area inundated by the Missoula floods. The line defining the area inundated by the Missoula floods in the Portland Basin was interpreted using multiple GIS datasets, including maximum flood contour lines defined by previous research in the area, mapped locations of glacial erratics and Missoula floods deposits, and lidar-derived images used to identify large-scale fluvial geomorphic features.

Courtesy of Oregon Department of Geology and Mineral Industries, 2012.

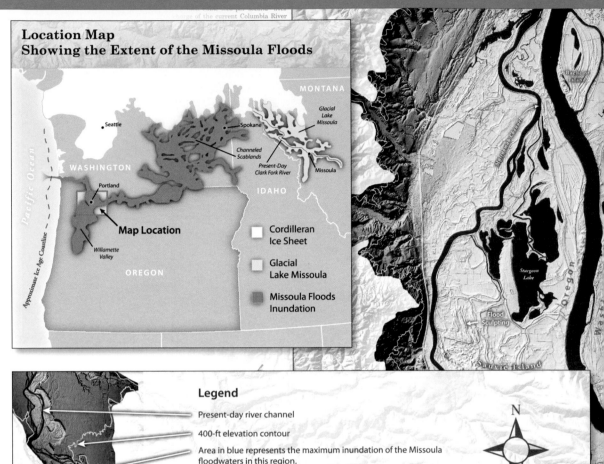

**Location Map
Showing the Extent of the Missoula Floods**

MONTANA

Seattle

Spokane

Glacial Lake Missoula

Channeled Scablands

Present-Day Clark Fork River

Missoula

WASHINGTON

Portland

IDAHO

Pacific Ocean

Map Location

Approximate Ice Age Coastline

Willamette Valley

OREGON

☐ Cordilleran Ice Sheet

☐ Glacial Lake Missoula

☐ Missoula Floods Inundation

Legend

— Present-day river channel

— 400-ft elevation contour

Area in blue represents the maximum inundation of the Missoula floodwaters in this region.

Area outside the blue extent was not directly affected by Missoula floodwaters.

N

0   5   10 Miles

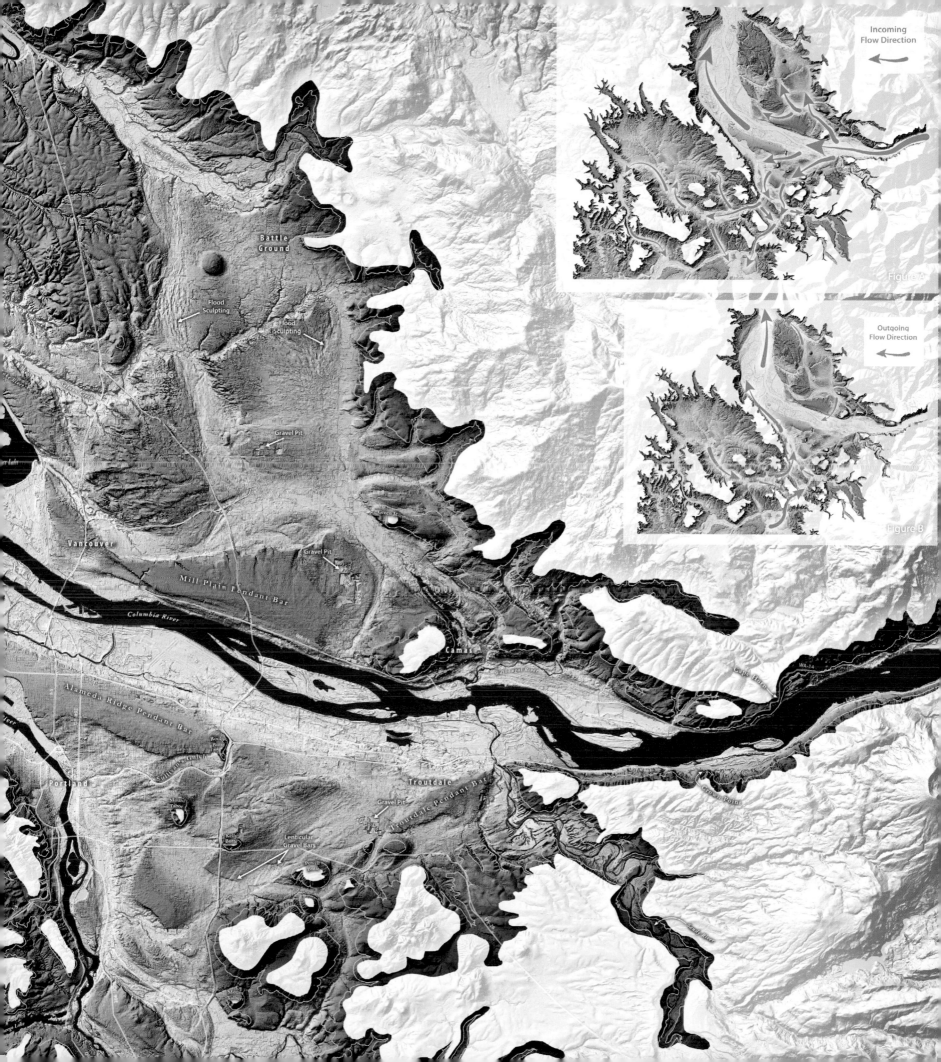

Figure A
Figure B

Incoming
Flow Direction

Outgoing
Flow Direction

Battle
Ground

Flood
Sculpting

Flood
Sculpting

Figure G

Gravel Pit

Green Mountain

Gravel Pit

Vancouver

Mill Plain Pendant Bar

Columbia River

Camas

Cape Horn

WA-14

Alameda Ridge Pendant Bar

Rocky
Butte

Sullivans Gulch

Portland

Troutdale

Troutdale Pendant Bar

Gravel Pit

Mount
Tabor

Kelly
Butte

Powell
Butte

Lenticular
Gravel Bars

Lidar Illuminated 2013 Calendar

Oregon Department of Geology and Mineral Industries
Portland, Oregon, USA
By Daniel Coe

Contact
Daniel Coe
dan.coe@dogami.state.or.us

Software
ArcGIS 10 for Desktop, Quick Terrain Modeler 7.0,
Adobe Photoshop CS5.1, Adobe Illustrator CS5.1

Data Sources
DOGAMI, National Agriculture Imagery Program, Oregon
Parks and Recreation Department

The Oregon Department of Geology and Mineral Industries (DOGAMI) uses lidar to create new-generation maps that are more accurate and comprehensive than any in the past. DOGAMI, via the Oregon Lidar Consortium, is continually acquiring new lidar data, which is being used to map and model geology and geologic hazards throughout Oregon.

The imagery shown here highlights the different ways lidar can be used to visualize Oregon's landscape, with particular attention to the state's outstanding geologic features. Innovative methods were used to display and combine bare-earth (last return), highest-hit (first return), canopy, and point cloud datasets to create perspective views of some of Oregon's well-known and lesser-known landforms. For images with continuous data, lidar was converted into 3-foot-resolution grids, which were layered and combined to form the final images. The point cloud imagery shows data with a resolution of at least 8 points per square meter. The lidar data in this product was collected and processed by Watershed Sciences Incorporated, Portland, Oregon.

DOGAMI produces maps and reports for the public and government to reduce the loss of life and property due to geologic hazards and to manage water and other geologic resources. DOGAMI helps Oregonians understand and prepare for earthquakes, tsunamis, coastal erosion, landslides, floods, and other geologic hazards.

Courtesy of Oregon Department of Geology and Mineral Industries, 2012.

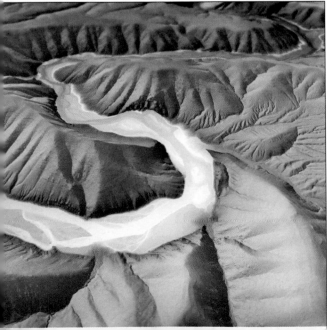

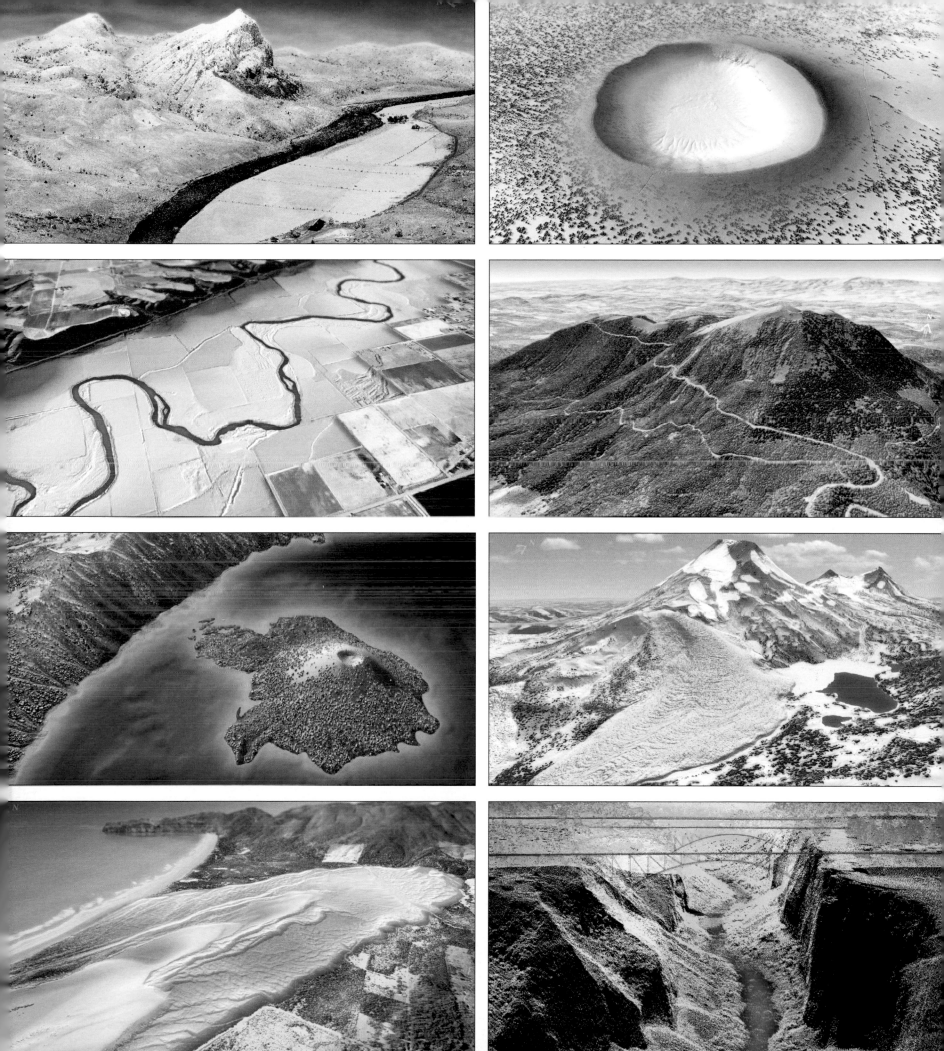

Regional Emergency Planning Maps

Fire Rescue Service of the Liberec Region
Liberec, Czech Republic
By Jan Petr and Jana Havrdova

Contact
Jana Havrdova
jana.havrdova@hzslk.cz

Software
ArcGIS 10 for Desktop

Data Sources
The Czech Office for Surveying, Mapping and Cadastre; Fire Rescue Service of the Liberec Region; Road and Motorway Directorate of the Czech Republic; Central European Data Agency; Water Research Institute, Liberec Region

Good planning is the basis of preparation for emergency events. Fire Rescue Service of the Liberec Region produces maps displaying objects that may compromise public safety and elements that protect people from possible threats (systems of early warning, areas intended for evacuation, etc.). Proper visualization and sufficiency of information ensure better and faster intervention of the Integrated Rescue System.

Among the potential sources of danger are places that store hazardous materials, livestock and poultry farms, watercourses, and areas with landslide risk. Long-term accident monitoring of roads and railways has resulted in the classification of dangerous sections. Based on population density and information about the potential dangers, places to install the emergency alert system with sirens to warn the population of danger were identified. Also identified were evacuation sites to ensure the essential needs of the population are met.

Courtesy of Fire Rescue Service of the Liberec Region.

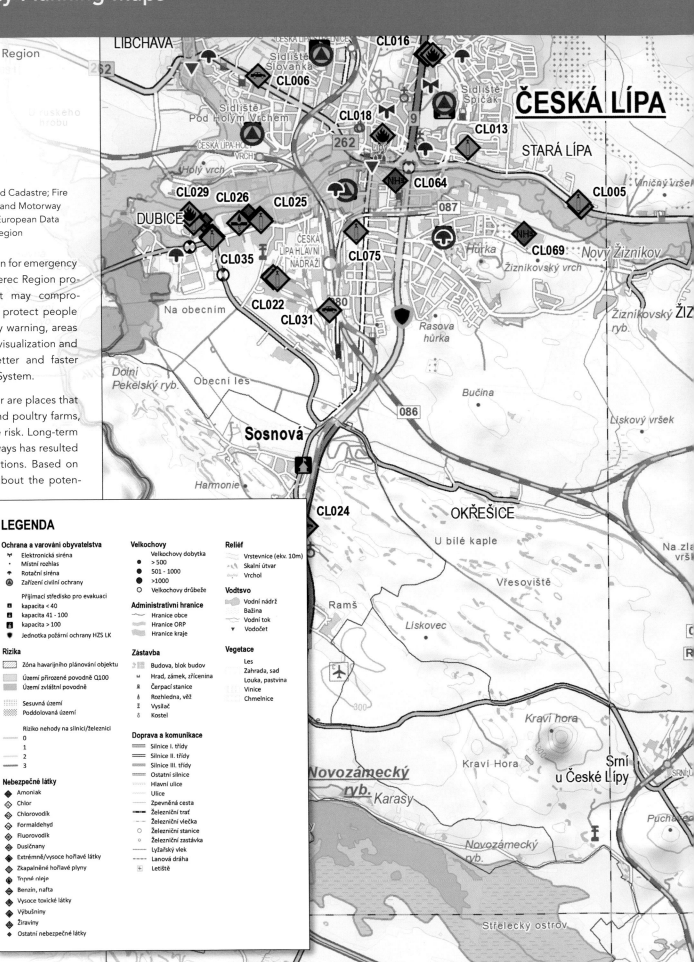

LEGENDA

Ochrana a varování obyvatelstva
- Elektronická siréna
- Místní rozhlas
- Rotační siréna
- Zařízení civilní ochrany

Přijímací středisko pro evakuaci
- kapacita < 40
- kapacita 41 - 100
- kapacita > 100
- Jednotka požární ochrany HZS LK

Rizika
- Zóna havarijního plánování objektu
- Území přirozené povodně Q100
- Území zvláštní povodně
- Sesuvná území
- Poddolovaná území

Riziko nehody na silnici/železnici
- 0
- 1
- 2
- 3

Nebezpečné látky
- Amoniak
- Chlor
- Chlorovodík
- Formaldehyd
- Fluorovodík
- Dusičnany
- Extrémně/vysoce hořlavé látky
- Zkapalněné hořlavé plyny
- Topné oleje
- Benzín, nafta
- Vysoce toxické látky
- Výbušniny
- Žíraviny
- Ostatní nebezpečné látky

Velkochovy
Velkochovy dobytka
- > 500
- 501 - 1000
- >1000
- Velkochovy drůbeže

Administrativní hranice
- Hranice obce
- Hranice ORP
- Hranice kraje

Zástavba
- Budova, blok budov
- Hrad, zámek, zřícenina
- Čerpací stanice
- Rozhledna, věž
- Vysílač
- Kostel

Doprava a komunikace
- Silnice I. třídy
- Silnice II. třídy
- Silnice III. třídy
- Ostatní silnice
- Hlavní ulice
- Ulice
- Zpevněná cesta
- Železniční trať
- Železniční vlečka
- Železniční stanice
- Železniční zastávka
- Lyžařský vlek
- Lanová dráha
- Letiště

Reliéf
- Vrstevnice (ekv. 10m)
- Skalní útvar
- Vrchol

Vodtsvo
- Vodní nádrž
- Bažina
- Vodní tok
- Vodočet

Vegetace
Les
- Zahrada, sad
- Louka, pastvina
- Vinice
- Chmelnice

Racial and Ethnic Diversity in Residential Patterns

GreenInfo Network
San Francisco, California, USA
By Alexandra Barnish and Larry Orman

Contact
Megan Dreger
megan@greeninfo.org

Software
ArcGIS for Desktop

Data Source
US Census

This map uses a unique approach to display Census 2010 population data. San Francisco County residential land uses were used to "cookie cutter" the census information. This eliminates neighborhoods dominated by commercial or industrial land uses and, therefore, shows a more accurate block-by-block footprint of where people live.

The population of San Francisco is racially and ethnically diverse. In 2010, Caucasian (non-Hispanic/Latino) and Asian residents comprised the majority of San Francisco's total population (76 percent). Data on race and ethnicity was derived from the 2010 decennial census. The Census Bureau collects this information by asking residents to choose the race(s) with which

they most closely identify and to indicate whether or not they are of Hispanic or Latino ethnicity. These self-identification questions include many categories that allow for numerous combinations. This makes the data complex to analyze and map.

The design of this map is intended to emphasize diversity by using colors with similar saturations. The green and purple colors chosen for the Hispanic/Latino and African American populations were intended to have a similar visual hierarchy although they are minorities within the city (23 percent). Neighborhood boundaries were overlaid for context. A coastal fade accents San Francisco's unique geography at the tip of a peninsula.

Courtesy of GreenInfo Network.

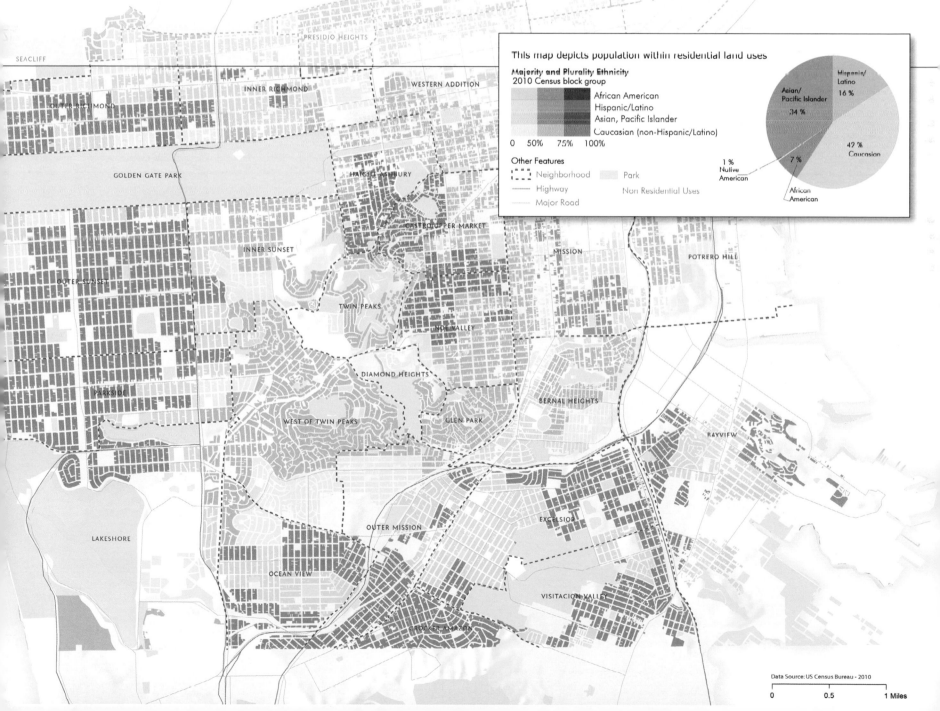

California Coastal and Marine Program—Protecting Fish Stocks and Livelihoods

The Nature Conservancy
San Francisco, California, USA
By Matt Merrifield

Contact
Matt Merrifield
mmerrifield@tnc.org

Software
ArcGIS 10 for Desktop

Data Sources
The Nature Conservancy, California Department of Fish and Game, National Marine Fisheries Service, Esri

The Nature Conservancy in California has pioneered a first-of-its-kind fishery reform approach that aligns community, fishing industry, and conservation interests to drive strategic changes in fishery management and harvest practices. The project lies within the California Current Large Marine Ecosystem, which is one of the four temperate upwelling systems in the world that support 20 percent of the world's commercial fish catch.

The productivity from upwelling of deep nutrient-rich waters, combined with the highly diverse marine habitats found within deepwater canyons, banks, seamounts, and expanses of soft bottom areas, supports a very high diversity of species. The Central Coast project area, approximately 9.6 million acres, spans the near-shore and offshore waters between Point Reyes and Point Conception and includes three national marine sanctuaries (Monterey Bay, Cordell Bank, and Gulf of the Farallons).

The Nature Conservancy has pursued conservation through private arrangements with fishermen along the Central Coast using a variety of gear types shown on the map. The organization uses a geographic information system to map and manage data related to fishing zones. Additionally, it collects geographic data on the location of fishing events and determines compliance with fishing plans. Geographic data on fishing events and fishing plans are made available to The Nature Conservancy's fishing partners through a web-based map application called eCatch.org, which provides real-time access to fishing data for trip planning and refinement of zones within the fishing plans.

Courtesy of The Nature Conservancy.

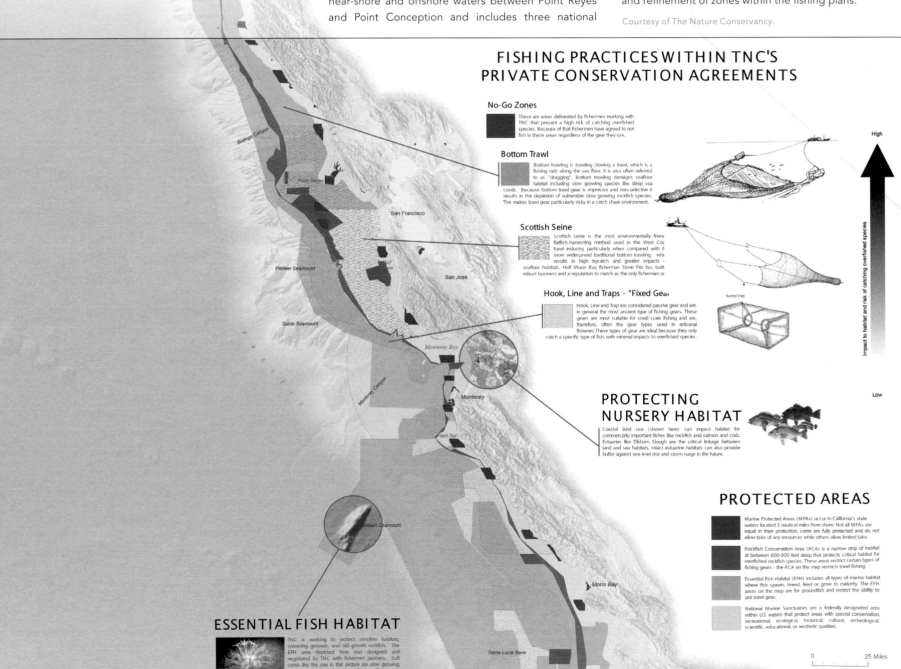

FISHING PRACTICES WITHIN TNC'S PRIVATE CONSERVATION AGREEMENTS

No-Go Zones
These are areas delineated by fishermen working with TNC that present a high risk of catching overfished species. Because of that fishermen have agreed to not fish in these areas regardless of the gear they use.

Bottom Trawl
Bottom trawling is trawling (towing a trawl, which is a fishing net) along the sea floor. It is also often referred to as "dragging". Bottom trawling damages seafloor habitat including slow growing species like deep sea corals. Because bottom trawl gear is imprecise and non-selective it results in the depletion of vulnerable slow growing rockfish species. This makes trawl gear particularly risky in a catch share environment.

Scottish Seine
Scottish seine is the most environmentally friendly flatfish-harvesting method used in the West Coast trawl industry, particularly when compared with the more widespread traditional bottom trawling, which results in high bycatch and greater impacts seafloor habitats. Half Moon Bay fisherman Steve Fitz has built robust business and a reputation to match as the only fisherman or

Hook, Line and Traps - "Fixed Gear"
Hook, Line and Trap are considered passive gear and are in general the most ancient type of fishing gears. These gears are most suitable for small scale fishing and are, therefore, often the gear types used in artisanal fisheries. These types of gear are ideal because they only catch a specific type of fish, with minimal impacts to overfished species.

PROTECTING NURSERY HABITAT
Coastal land use (shown here) can impact habitat for commercially important fishes like rockfish and salmon and crab. Estuaries like Elkhorn Slough are the critical linkage between land and sea habitats. Intact estuarine habitats can also provide buffer against sea-level rise and storm surge in the future.

PROTECTED AREAS
Marine Protected Areas (MPAs) occur in California's state waters located 3 nautical miles from shore. Not all MPAs are equal in their protection, some are fully protected and do not allow take of any resources while others allow limited take.

Rockfish Conservation Area (RCA) is a narrow strip of habitat at between 600-900 feet deep that protects critical habitat for overfished rockfish species. These areas restrict certain types of fishing gears - the RCA on this map restricts trawl fishing.

Essential Fish Habitat (EFH) includes all types of marine habitat where fish spawn, breed, feed or grow to maturity. The EFH areas on the map are for groundfish and restrict the ability to use trawl gear.

National Marine Sanctuaries are a federally designated area within US waters that protect areas with special conservation, recreational, ecological, historical, cultural, archeological, scientific, educational, or aesthetic qualities.

ESSENTIAL FISH HABITAT
TNC is working to protect sensitive habitats, spawning grounds, and old growth rockfish. The EFH area depicted here was designed and negotiated by TNC with fishermen partners. Soft corals like the one in this picture are slow growing and vulnerable to bottom trawl fishing. Seamounts like Davidson are unique in the marine environment and provide critical habitat for corals and rockfish.

Mapping Change within Sundarband Mangrove Forest

US Department of Agriculture Forest Service, Fire and Aviation Management Sacramento
Sacramento, California, USA
By Tanushree Biswas, Paul Maus, Kevin Megown, and Linda Smith (graphic design)

Contact
Tanushree Biswas
tbiswas@fs.fed.us

Software
ArcGIS Desktop 9.3

Data Sources
Landsat TM 1989, 1999, and 2009

The Sundarban Reserve Forest (SRF) is the largest contiguous mangrove forest in the world. Forty percent of the SRF is managed by India's West Bengal Forest Department. The mangrove forest makes up 60 percent of Bangladesh's natural forest. These areas require more advanced monitoring than current field methods provide. The US Forest Service Remote Sensing Applications Center developed remote-sensing-based methods and provided training to the Bangladesh Forest Department's Resource Information Management System (RIMS) unit to efficiently monitor change in the mangroves of the SRF. This effort supports the United Nations' collaborative program on Reducing Emissions from Deforestation and Forest Degradation in Developing Countries (REDD+) initiatives within Bangladesh.

Courtesy of US Department of Agriculture Forest Service Remote Sensing Applications Center.

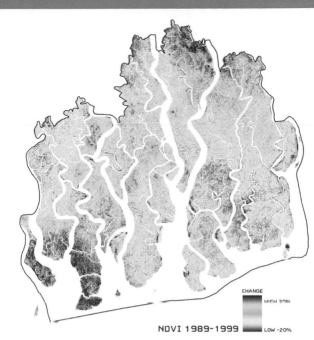

NDVI 1989-1999

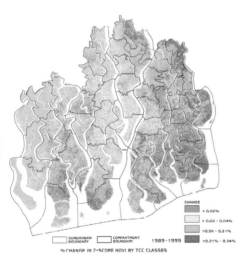

% CHANGE IN Z-SCORE NDVI BY TCC CLASSES

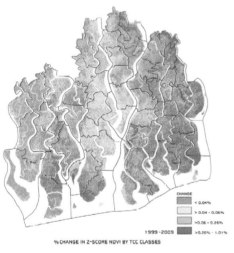

% CHANGE IN Z-SCORE NDVI BY TCC CLASSES

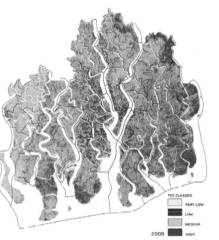

TCC CLASSES DERIVED FROM NDVI

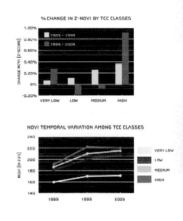

MANAGEMENT COMPARTMENT: Percent changes in z-NDVI reduced from ± 4 % (1989-1999) to ± 2% (1999-2009), indicating an increase in forest cover. This analysis assisted in identifying potential areas for management. These results were in agreement with the independent field study for biomass and carbon stock estimation conducted by the USFS in April 2010. Field plots that showed a decline in carbon stock between 1995 and 2010 were found to be nested within the management units or region that showed a decline in NDVI from the z-score change analysis. Agreement among these independent studies confirmed the the z-score results. This suggests that the pattern observed at a plot level is an indication of disturbance and carbon stock loss by the z-score analysis at the landscape level.

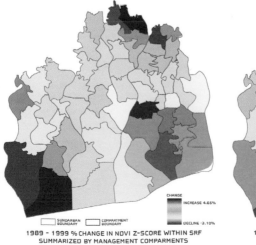

1989 - 1999 % CHANGE IN NDVI Z-SCORE WITHIN SRF SUMMARIZED BY MANAGEMENT COMPARMENTS

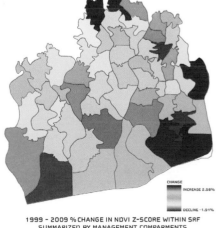

1999 - 2009 % CHANGE IN NDVI Z-SCORE WITHIN SRF SUMMARIZED BY MANAGEMENT COMPARMENTS

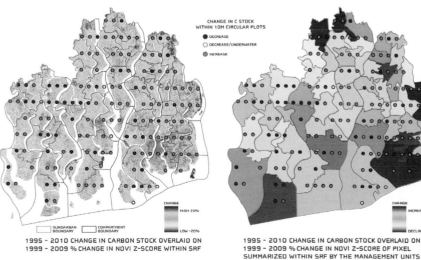

1995 - 2010 CHANGE IN CARBON STOCK OVERLAID ON 1999 - 2009 % CHANGE IN NDVI Z-SCORE WITHIN SRF

1995 - 2010 CHANGE IN CARBON STOCK OVERLAID ON 1999 - 2009 % CHANGE IN NDVI Z-SCORE OF PIXEL SUMMARIZED WITHIN SRF BY THE MANAGEMENT UNITS

Energy Hogs of New York City

Columbia University, Modi Research Group
New York, New York, USA
By Shaky Sherpa, Bianca Howard, and Vijay Modi

Contact
Shaky Sherpa
ss3491@columbia.edu

Software
ArcGIS 10 for Desktop

Data Source
New York City Department of City Planning, New York
City Office of Long-Term Planning; Residential Energy
Consumption Survey; Commercial Buildings Energy
Consumption Survey

The map represents an estimate of the total annual building energy consumption at the block level and at the tax lot level for New York City and is expressed in kilowatt hours (kWh) per square meter of land area. The annual building energy consumption was estimated using ZIP code-level energy usage on electricity, natural gas, fuel oil, and steam consumption for 2009 as well as building information obtained from MapPLUTO, a New York City Department of City Planning geographic database.

Through statistical regression, annual energy usage intensities were estimated. Energy usage intensity (EUI), is annual energy consumption divided by the total building floor area. These are delivered energy intensities and not primary energy intensities. This distinction is critical since primary energy utilized to produce electricity can vary with the type of power plant. In addition, residential multi-family estimates for Manhattan and Bronx are specific to those boroughs. Office estimates for Manhattan are specific only to Manhattan. Energy intensities were first estimated for electricity and for all fuels, including steam supply.

Data from the residential energy consumption survey (RECS) and the commercial building energy consumption survey (CBECS) was used to break the energy use down into specific end uses: space heating, space cooling, water heating, and base electric applications (lighting and plug loads).

Courtesy of Modi Research Group, 2012.

Energy consumption at tax lot level

(kWh/METER² OF BLOCK AREA)

N/A	<=50	75	90	120	170	200	300	400	600	900	1200	1750	2500	5000	>5000

A 100-watt lightbulb turned on for ten hours uses one kilowatt-hour (kWh) of energy.

Woods Hole Research Center
Falmouth, Massachusetts, USA
By Paul A. Lefebvre Jr., Leandro Castello, Michael Coe,
and Laura Hess

Contact
Paul Lefebvre
paul@whrc.org

Software
ArcGIS 10.1 for Desktop

Data Sources
NASA, International Rivers, Agencia Nacional de Águas
(Brazil), IBAMA (Brazil), Conservation International, IPAM
(Brazil)

The hydrological connectivity of freshwater ecosystems in the Amazon basin makes them highly sensitive to a broad range of anthropogenic activities occurring in aquatic and terrestrial systems at local and distant locations. Current management policies and the existing network of protected areas focus on terrestrial ecosystems, and fail to account for the hydrologic connectivity of freshwater ecosystems. As a result, the ability of those systems to provide historically important goods and services is declining. Deforestation in headwaters landscapes alters the timing, volume, temperature, and quality of runoff, impacting aquatic biodiversity. Channelization and dam building on larger rivers affects the timing and volume of floodwaters while also impeding fish migration. Municipal pollution directly contaminates large stretches of river, and petroleum exploration and production in far upstream regions places all downstream aquatic systems at risk in the event of a spill.

Economic pressures continue to push agricultural expansion, and improvements to the transportation infrastructure are being completed to connect this new breadbasket to both European and Asian markets. It is time to develop a river catchment-based conservation framework for the whole basin that protects both aquatic and terrestrial ecosystems.

This map was originally developed in support of an article titled "The Vulnerability of Amazon Freshwater Ecosystems" in the journal *Conservation Letters*.

Courtesy of Woods Hole Research Center.

Legend

	Deforested
	Deforested - Pasture
	Deforested - Soy
	Reserve
	Seasonally Flooded
	Petroleum lease
	Petroleum proposed
	Deepwater Port
	Unimproved Road
	Paved Road
	Newly Paved Road
	Maintained Waterway
	Dammed Rivers

Hydroelectric Dam
Capacity Mw

- 2501 - 5000
- 5001 - 7500
- 7501 - 10000
- 10001 - 11233

Status
- Planned
- Construction
- Operational

0   62.5   125   250   375   500 Km

Guiana Shield

Macapá

Barcelos

Belém

Tefé

Manaus   Itacoatiara

Coari

Santarém

Iquitos

Humaitá

Brazilian Shield

Branco

Bananal

Llanos de Moxos

Brasília

Goiânia

Peru

Brazil

Identifying Floodplain Restoration Opportunities in California's Central Valley

AECOM
Sacramento, California, USA
By F. Eryn Pimentel

Contact
Eryn Pimentel
eryn.pimentel@aecom.com

Software
ArcGIS 10 for Desktop, HEC-RAS

Data Sources
California Department of Conservation, California Department of Fish and Game, California Department of Transportation, California Department of Water Resources, California Energy Commission, California Environmental Protection Agency, GreenInfo Network

Floodplain inundation potential (FIP) and other opportunities and constraints were used along the Sacramento and San Joaquin Rivers and their major tributaries to support restoration planning for the conservation strategy of the Central Valley Flood Protection Plan for the California Department of Water Resources.

Physical suitability was identified using lidar and hydraulic modeling software. Fifty percent and 67 percent sustained chance floods were used as surrogates for reasonably frequent floodplain activation flows that may benefit several sensitive endemic fish and wildlife species. Identifying constraints involved mapping land cover within the 2-mile-wide modeling extent, as well as overlaying existing conserved areas and major infrastructure.

Of particular interest as potential restoration opportunity areas were those with greater potential for floodplain inundation that are isolated from rivers by levees whose physical condition is of high concern. These areas could provide opportunities to reconnect the river with its floodplain, enhancing floodplain and riverine ecosystems. Riparian and wetland habitat and conserved areas also provide distinct types of restoration opportunities.

Results show that extensive areas with restoration potential exist, both connected and disconnected from rivers, and also that using FIP is an effective concept for regional planning. This analysis aids in developing a strategic vision for California's Central Valley to reduce the risk of flooding while creating and improving unique natural habitat

Special thanks to contributors Stacy Cepello, Kevin G. Coulton, Lee D. von Gynz-Guethle, John C. Hunter, Ray McDowell, and Jonathan D. McLandrich. Courtesy of AECOM.

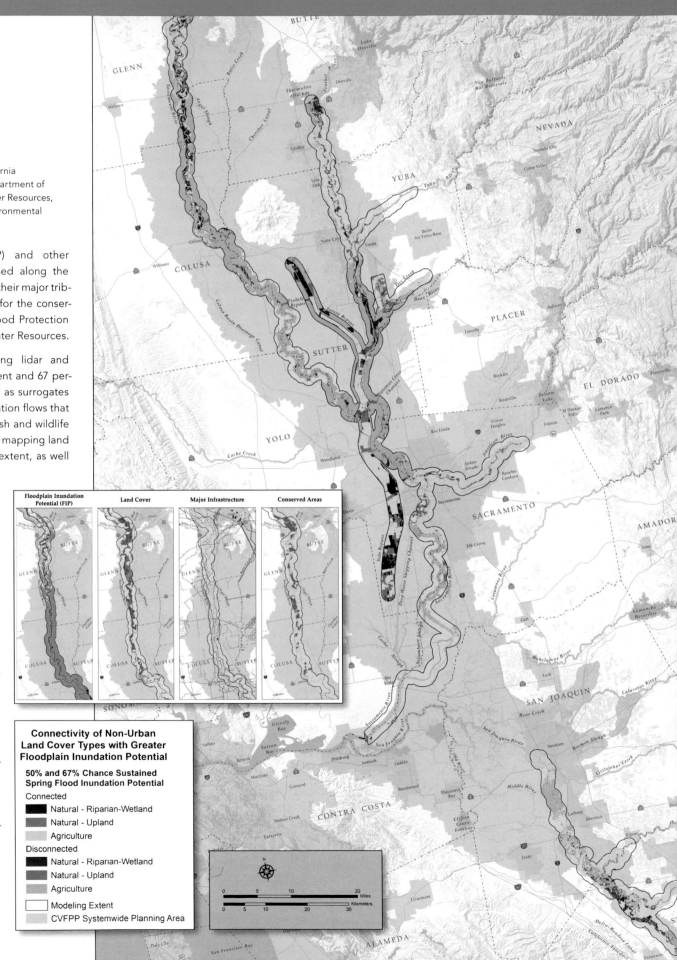

Connectivity of Non-Urban Land Cover Types with Greater Floodplain Inundation Potential

50% and 67% Chance Sustained Spring Flood Inundation Potential

Connected
- Natural - Riparian-Wetland
- Natural - Upland
- Agriculture

Disconnected
- Natural - Riparian-Wetland
- Natural - Upland
- Agriculture

- Modeling Extent
- CVFPP Systemwide Planning Area

Location of Course Dives in the Canyons of Lacaze-Duthiers Bourcart and the Rocks of Sete

GIS Posidonie and Agence des Aires Marines Protegées (GIS Posidonia and Agency of the Marine Protected Areas)
Marseille, Provence-Alpes-Côte d'Azur, France
By Adrien Goujard

Contact
Adrien Goujard
adrien.goujard@univ-amu.fr

Software
ArcGIS 10 for Desktop, ArcGIS 3D Analyst, ArcGIS Spatial Analyst

Data Source
Agence des Aires Marines Protegées

MEDSEACAN was a vast campaign carried out by the GIS Posidonie and Agence des Aires Marines Protegées between November 2008 and April 2010 with the goal of exploring the heads of continental Mediterranean canyons. This map illustrates the effort of dives carried out from the underwater canyon of Lacaze-Duthiers to the canyon of Bourcart.

The dives were distributed around the canyons to facilitate their intercomparison. This sector comprises thirty-five dives covering nearly 42,770 meters over seventy-one hours of dives in a remotely operated vehicle (ROV) and a submarine.

From the navigation records of the ROV and the submarine, the various observations (species, waste, traces of trawl, nature of the substrate, etc.) carried out during the prospecting could be geolocated.

Courtesy of GIS Posidonie and Agence des Aires Marines Protegées.

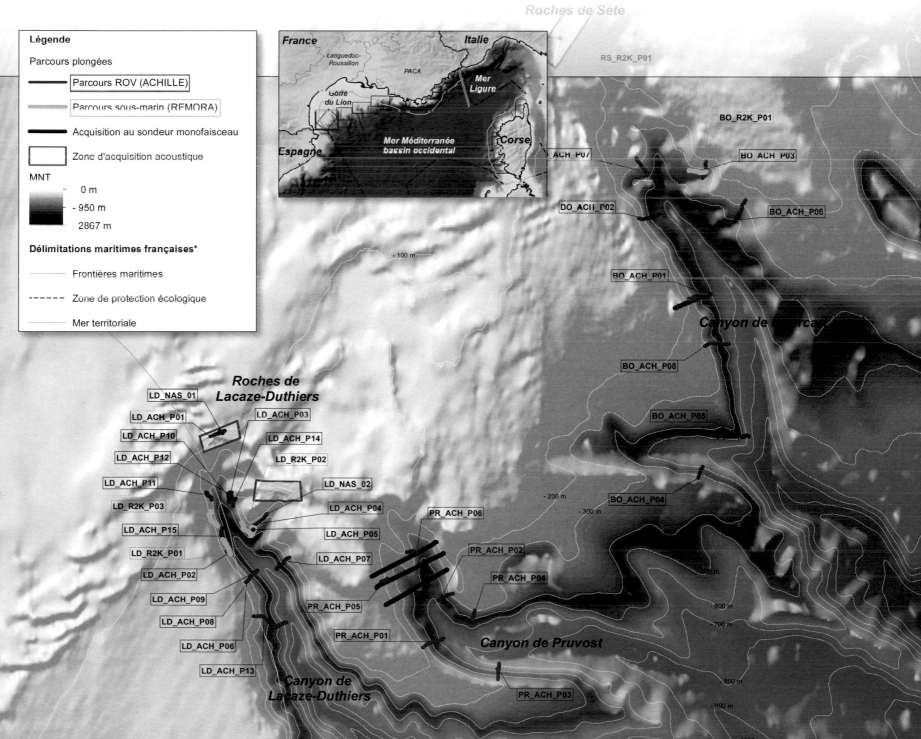

Archaeological Prediction for Road Asset Management

Opus International Consultants Ltd
Hamilton, Waikato, New Zealand
By Andrew Standley and Nick Cable

Contact
Andrew Standley
andrew.standley@opus.co.nz

Software
ArcGIS 10 for Desktop

Data Sources
Institute of Geological and Nuclear Sciences, Far North District Council, Land Information New Zealand, Northland Regional Council, New Zealand Archaeological Association, Historic Places Trust

This innovative project for the New Zealand Transport Agency involved the development of a GIS model to identify areas of potential archaeological risk to the state highway network. A deductive site suitability approach was taken using ArcGIS fuzzy logic techniques. This expert system was guided by an archaeologist with extensive knowledge of the State Highway 11 study area on the Northland east coast. The deductive model hypothesis focused on environmental influencing factors that are thought to have contributed to human settlement decision making prior to European contact.

The model design involved the following:

1. Preference for volcanic rocks (i.e., for gardening, settlement foci around scoria cones)

2. Preference for low elevation and flat slope using walking time path distance catchments

3. Preference for proximity to coastline, lakes, and rivers

A user guide table was developed to help interpret the archaeological risk map and indicate when archaeologists should be engaged for particular types of planned roadwork. The model was tested using New Zealand Archaeological Association site data resulting in a model accuracy of 81 percent. Recommended future enhancements include social and cultural influences on human settlement patterns, particularly the identification of regional travel corridors.

Courtesy of Opus International Consultants Ltd.

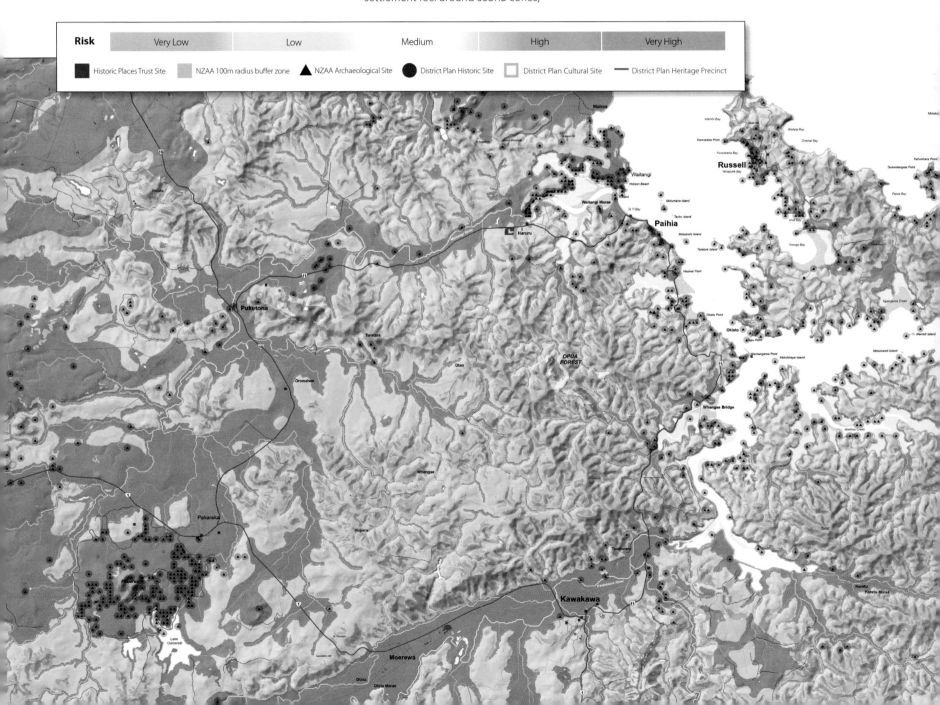

Amazon 2012—Protected Areas and Indigenous Territories

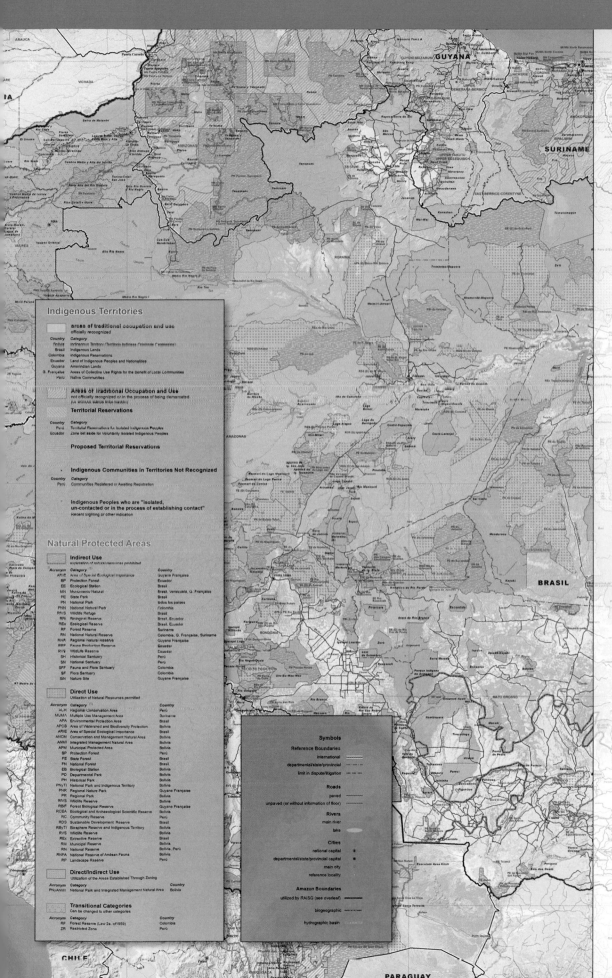

Indigenous Territories

areas of traditional occupation and use
officially recognized

Country	Category
Bolivia	Indigenous Territory (Territorio Indígena Originario Campesino)
Brasil	Indigenous Lands
Colombia	Indigenous Reservations
Ecuador	Land of Indigenous Peoples and Nationalities
Guyana	Amerindian Lands
G. Française	Areas of Collective Use Rights for the Benefit of Local Communities
Perú	Native Communities

Areas of Traditional Occupation and Use
not officially recognized or in the process of being demarcated
(or without status information)

Territorial Reservations

Country	Category
Perú	Territorial Reservations for Isolated Indigenous Peoples
Ecuador	Zone set aside for Voluntarily Isolated Indigenous Peoples

Proposed Territorial Reservations

Indigenous Communities in Territories Not Recognized

Country	Category
Perú	Communities Registered or Awaiting Registration

Indigenous Peoples who are "isolated, un-contacted or in the process of establishing contact"
Recent Sighting or other indication

Natural Protected Areas

Indirect Use
exploitation of natural resources prohibited

Acronym	Category	Country
ARJE	Area of Special Ecological Importance	Guyane Française
BP	Protection Forest	Ecuador
EE	Ecological Station	Brasil
MN	Monumento Natural	Brasil, Venezuela, G. Française
PE	State Park	Brasil
PN	National Park	todos los países
PNN	National Natural Park	Colombia
RVS	Wildlife Refuge	Brasil
RN	Biological Reserve	Brasil, Ecuador
REe	Ecological Reserve	Brasil, Ecuador
RF	Forest Reserve	Suriname
RN	National Natural Reserve	Colombia, G. Française, Suriname
RNA	Regional Natural Reserve	Guyane Française
RPF	Fauna Production Reserve	Ecuador
RVS	Wildlife Reserve	Ecuador
SH	Historical Sanctuary	Perú
SN	National Sanctuary	Perú
SFF	Fauna and Flora Sanctuary	Colombia
SF	Flora Sanctuary	Colombia
SN	Nature Site	Guyane Française

Direct Use
Utilization of Natural Resources permitted

Acronym	Category	Country
ACR	Regional Conservation Area	Perú
MUMA	Multiple Use Management Area	Suriname
APA	Environmental Protection Area	Brasil
APCB	Area of Watershed and Biodiversity Protection	Bolivia
ARIE	Area of Special Ecological Importance	Brasil
ANCM	Conservation and Management Natural Area	Bolivia
ANMI	Integrated Management Natural Area	Bolivia
APM	Municipal Protected Area	Bolivia
BP	Protection Forest	Bolivia
FE	State Forest	Brasil
FN	National Forest	Brasil
EB	Biological Station	Bolivia
PD	Departmental Park	Bolivia
PH	Historical Park	Bolivia
PNyTI	National Park and Indigenous Territory	Bolivia
PNR	Regional Nature Park	Guyane Française
PR	Regional Park	Bolivia
RVS	Wildlife Reserve	Bolivia
RBF	Forest Biological Reserve	Guyane Française
RCEA	Ecological and Archaeological Scientific Reserve	Bolivia
RC	Community Reserve	Perú
RDS	Sustainable Development Reserve	Brasil
RByTI	Biosphere Reserve and Indigenous Territory	Bolivia
RVS	Wildlife Reserve	Bolivia
REe	Extractive Reserve	Brasil
RM	Municipal Reserve	Bolivia
RN	National Reserve	Bolivia, Perú
RNFA	National Reserve of Andean Fauna	Bolivia
RP	Landscape Reserve	Perú

Direct/Indirect Use
Utilization of the Areas Established Through Zoning

Acronym	Category	Country
PNyANMI	National Park and Integrated Management Natural Area	Bolivia

Transitional Categories
Can be changed to other categories

Acronym	Category	Country
RF	Forest Reserve (Law 2a. of 1959)	Colombia
ZR	Restricted Zone	Perú

Symbols

Reference Boundaries
- international
- departmental/state/provincial
- limit in dispute/litigation

Roads
- paved
- unpaved (or without information of floor)

Rivers
- main river
- lake

Cities
- national capital
- departmental/state/provincial capital
- main city
- reference locality

Amazon Boundaries
- utilized by RAISG (see overleaf)
- biogeographic
- hydrographic basin

Amazon Georeferenced Socio-Environmental Information Network

Bogota, Cundinamarca, Colombia
By Amazon Georeferenced Socio-Environmental Information Network

Contact
Milton Hernán Romero Ruiz
mromero@gaiaamazonas.org

Software
ArcGIS Desktop 9.3

Data Sources
Bolivia (SERNAP, Viceministerio de Tierras); Brazil (ISA); Colombia (IGAC, UAESPNN, IDEAM); Ecuador (Ecociencia, Ecolex, MAE); Guyane Francaise (DEAL); Guyana (DCW); Peru (IIAP, MINAM, SICNA); Suriname (DCW, WDPA); Venezuela (IVIC, Simon Bolivar Geographic)

The Amazon Georeferenced Socio-Environmental Information Network (RAISG) is a space for the exchange and networking of GIS-based socioenvironmental information. That information supports processes that actively link collective rights to the promotion and sustainability of the socioenvironmental diversity of the Amazon region. The main objective of RAISG is to facilitate cooperation among institutions that already use socioenvironmental geographic information systems in the Amazon region through a methodology that coordinates collective efforts through an accumulative, decentralized, and public process of information sharing.

Amazonia covers approximately 7.8 million square kilometers shared by nine countries. There are 33 million inhabitants and 385 indigenous groups. The protection of socioenvironmental diversity has been consolidated through the recognition of the indigenous territories (IT) and the constitution of protected natural areas (PNA). This has occurred both in the Brazilian Amazon (Brazil), as well as in the Guiana (Venezuela, Suriname, Guyana, and French Guiana) and Andean Amazon (Colombia, Ecuador, Peru, and Bolivia). Today, this area covers 3,502,750 square kilometers (2,144,412 square kilometers of IT and 1,696,529 square kilometers of PNA, with an overlap 336,365 square kilometers), which corresponds to 45 percent of the area. These areas are being transformed into forest islands surrounded by extractives activities (gas and oil, mining, and logging), large infrastructure (roads and hydroelectric plants), and agricultural projects (cattle ranching, plantations of biofuel crops or soya).

Courtesy of Amazon Georeferenced Socio-Environmental Information Network, 2012, and Gaia Amazonas Foundation.

Areas of Potential Climate Resiliency

GreenInfo Network
San Francisco, California, USA
By Alexandra Barnish

Contact
Megan Dreger
megan@greeninfo.org

Software
ArcGIS for Desktop

Data Source
UC Davis Hopland Research and Extension Center, Blueprint Team

This map is part of an analysis of biodiversity and climate change resiliency published in *Conservation Blueprint* by the Land Trust of Santa Cruz County. The map identifies areas that could facilitate species migration in response to the predicted hotter, drier climate of the future. These include existing wet areas, north-facing slopes with cooler microclimates, and contiguous habitat patches with high elevation range that can ease movement in response to a changing climate. The report calls these areas climate change refugia, which include areas that are wetter and cooler at present and generally scattered throughout the county. The report predicts wet areas will also be critical to human adaptation to climate change, so protecting intact habitat where species can migrate is another way to add resiliency.

The design of the map was intended to evoke the sense of the cooler, wetter areas by using cool and earth-friendly colors. In particular, the north-facing slopes were depicted as a cool shade of blue. Habitat patches with the highest range of elevation were emphasized using the map's most vibrant color. Masking areas outside Santa Cruz County makes the planning area "pop" while still providing context.

Courtesy of GreenInfo Network.

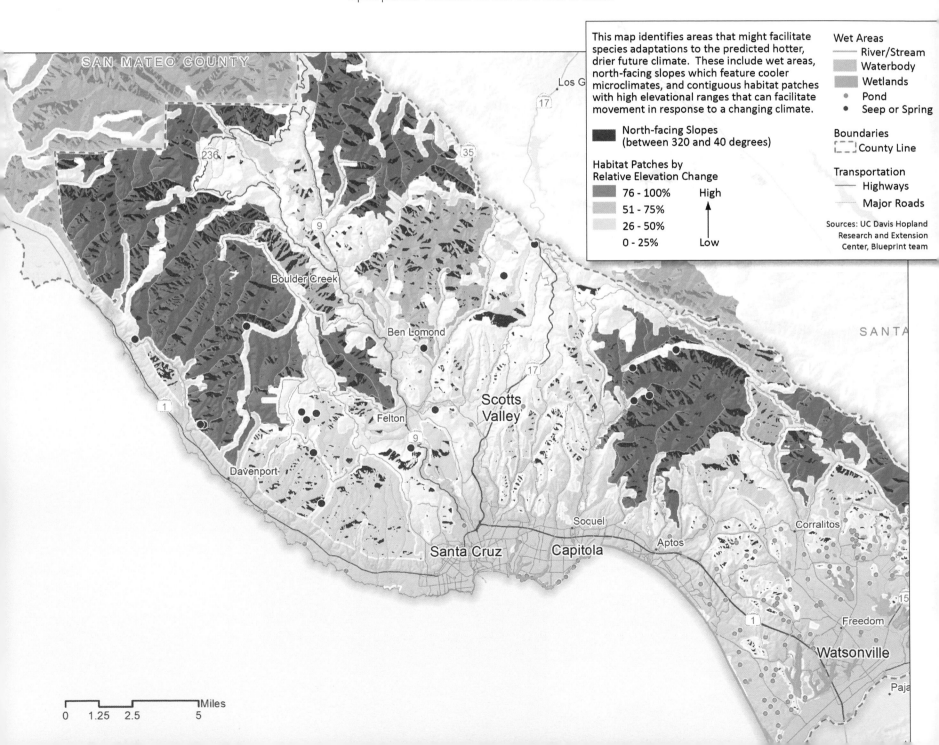

This map identifies areas that might facilitate species adaptations to the predicted hotter, drier future climate. These include wet areas, north-facing slopes which feature cooler microclimates, and contiguous habitat patches with high elevational ranges that can facilitate movement in response to a changing climate.

North-facing Slopes
(between 320 and 40 degrees)

Habitat Patches by Relative Elevation Change
76 - 100%   High
51 - 75%
26 - 50%
0 - 25%   Low

Wet Areas
— River/Stream
  Waterbody
  Wetlands
• Pond
• Seep or Spring

Boundaries
[- - -] County Line

Transportation
— Highways
— Major Roads

Sources: UC Davis Hopland Research and Extension Center, Blueprint team

SAN MATEO COUNTY

Los G

Boulder Creek

Ben Lomond

Scotts Valley

Felton

Davenport

Soquel

Aptos

Corralitos

Santa Cruz   Capitola

Freedom

Watsonville

SANTA

Paja

Miles
0   1.25   2.5   5

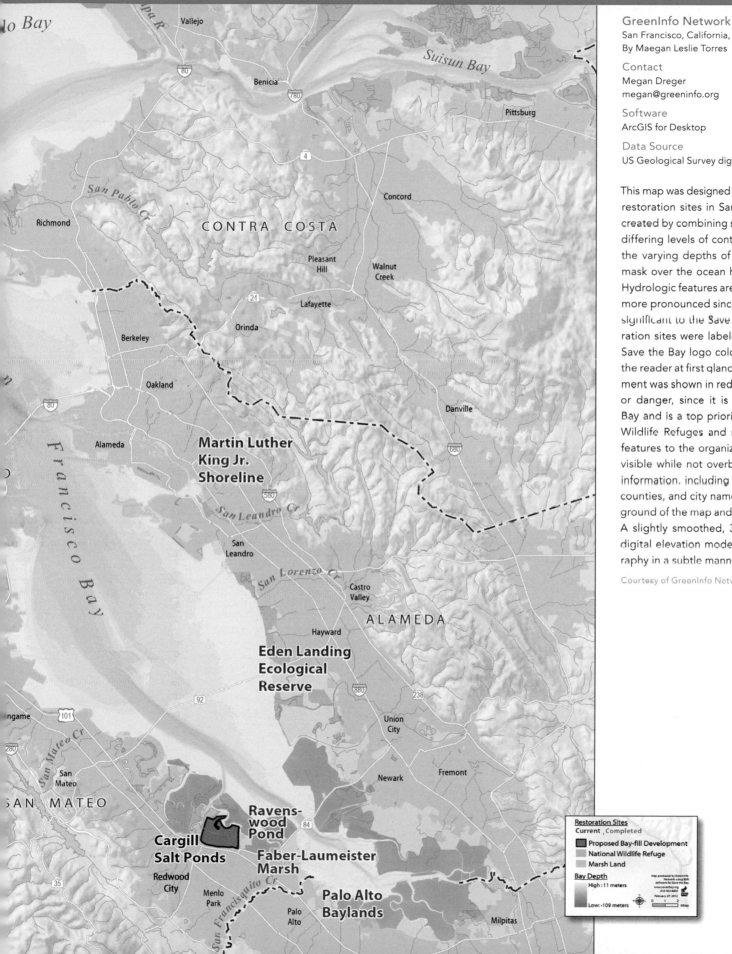

GreenInfo Network
San Francisco, California, USA
By Maegan Leslie Torres

Contact
Megan Dreger
megan@greeninfo.org

Software
ArcGIS for Desktop

Data Source
US Geological Survey digital elevation model

This map was designed as part of a brochure highlighting restoration sites in San Francisco Bay. The effect was created by combining several layers of bathymetry with differing levels of contrast and brightness to highlight the varying depths of the bay. A gradual transparent mask over the ocean helps draw the focus to the bay. Hydrologic features are labeled in a darker blue and are more pronounced since they drain into the bay and are significant to the Save the Bay organization. The restoration sites were labeled in bold text that reflects the Save the Bay logo colors. The sites should pop out to the reader at first glance. The proposed bay fill development was shown in red, indicating the sense of a threat or danger, since it is strongly opposed by Save the Bay and is a top priority of the organization. National Wildlife Refuges and marshland were also important features to the organization and were designed to be visible while not overbearing. All additional basemap information, including streams, highways, urban areas, counties, and city names, is designed to sit in the background of the map and serve solely as spatial reference. A slightly smoothed, 30-meter US Geological Survey digital elevation model was used to show land topography in a subtle manner.

Courtesy of GreenInfo Network.

Fort Sill Cantonment Area

US Army FIRES Center of Excellence and Fort Sill

Fort Sill, Oklahoma, USA
By Aaron Peterson

Contact
Aaron Peterson
aaron.e.peterson.civ@mail.mil

Software
ArcGIS 10 for Desktop

Data Source
US Army FIRES Center of Excellence and Fort Sill

The Directorate of Public Works, Geospatial Information and Services Office, at the US Army FIRES Center of Excellence (FCoE) and Fort Sill supports a wide variety of installation functions, from providing standardized field and air defense artillery targeting products to assisting with local emergency response. This office performs complex geospatial analysis using one of the largest geospatial repositories of the US Army.

The *Fort Sill Cantonment Area* map was created as a standardized 1:8,000-scale cartographic product to depict the real property infrastructure and emerging development of the sprawling installation. Decisions by the Defense Base Closure and Realignment Commission resulted in growth of nearly $1 billion in construction and modernization of the fort infrastructure. The installation needed a product to help the military and civilian clients track the projects and navigate the cantonment. The map is composed of 1-meter hillshade imagery and over fifty features that are created and maintained on the installation enterprise geospatial database.

Courtesy of US Army FIRES Center of Excellence and Fort Sill.

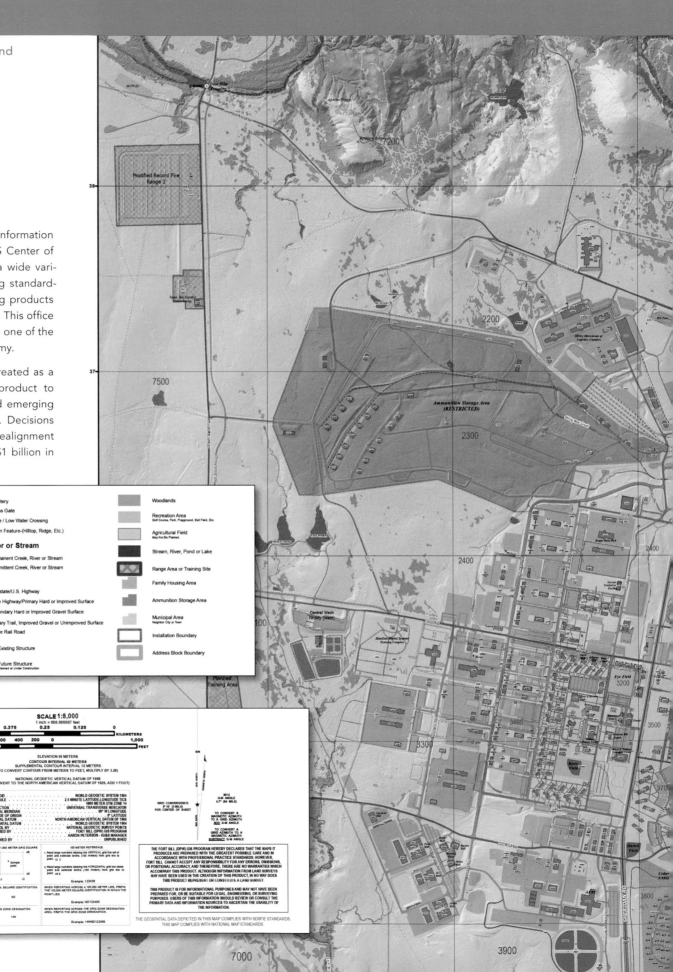

Distribution of CCBC Students by Degree-Seeking Classification

Community College of Baltimore County
Catonsville, Maryland, USA
By Alan Eyler

Contact
Scott Jeffrey
sjeffrey@ccbcmd.edu

Software
ArcGIS 10 for Desktop

Data Sources
CCBC Department of Planning, Research, and Evaluation

Community College of Baltimore County (CCBC) is using GIS to map the distribution of its students by degree-seeking classification in Central Maryland. Based on data provided by the CCBC Department of Planning, Research, and Evaluation (PRE), students were mapped based on whether they were seeking a degree or pursuing continuing education (noncredit/nondegree). To create this map, the PRE student database was geocoded to provide point locations based on each student's address.

The pattern of distribution is similar between credit-seeking students and noncredit-seeking students and indicates a higher concentration of CCBC student enrollment in areas of the southeast and southwest portions of Baltimore County, Maryland, where main campuses are located. In addition, Baltimore City, Maryland, shows a fairly even distribution of credit-seeking students, as well as noncredit-seeking students. The map provided the college administration with a visual picture of the student population and categories and is being used for targeted marketing and allocation of college resources.

Courtesy of Community College of Baltimore County Geospatial Applications Program.

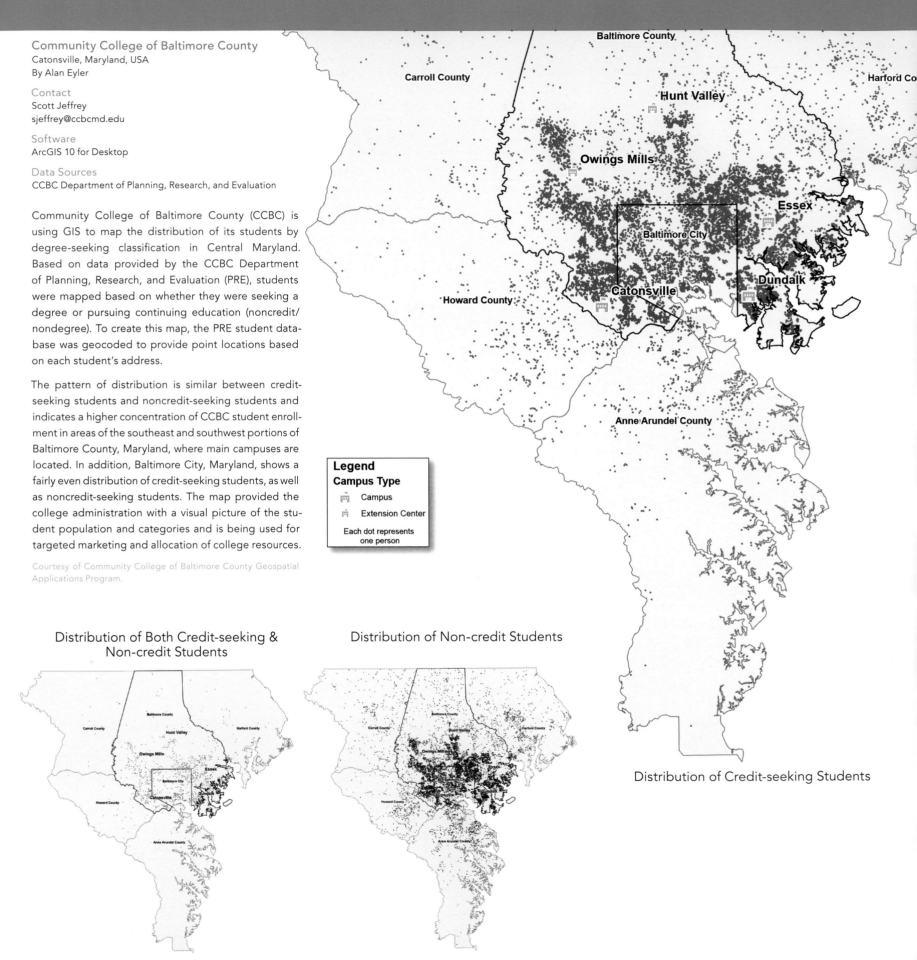

Legend
Campus Type
Campus
Extension Center
Each dot represents one person

Distribution of Both Credit-seeking & Non-credit Students

Distribution of Non-credit Students

Distribution of Credit-seeking Students

University of the Arctic Thematic Networks

Nordpil
Stockholm, Sweden
By Hugo Ahlenius (Nordpil) and Veli-Pekka Laitinen (University of the Arctic)

Contact
Hugo Ahlenius
hugo.ahlenius@nordpil.com

Software
ArcGIS for Desktop, Microsoft Access, Adobe Illustrator

Data Source
University of the Arctic

The University of the Arctic (UArctic) secretariat wanted to present its thematic networks on a map. UArctic is a network of higher education and research institutions in the northern hemisphere collaborating in exchange programs, research projects, and educational initiatives. Under UArctic are a number of thematic networks where the institutions participate. The goal of the map was not so much to present detail about the networks but more to show the abundance and activity of the members as a "ball of yarn" across the North Pole. Data was collected, prepared, and analyzed using Microsoft Access, and the map was put together in ArcGIS. Final edits and design were done in Adobe Illustrator. The background map was delivered by the client, designer Veli-Pekka Laitinen.

Courtesy of Hugo Ahlenius and Veli-Pekka Laitinen.

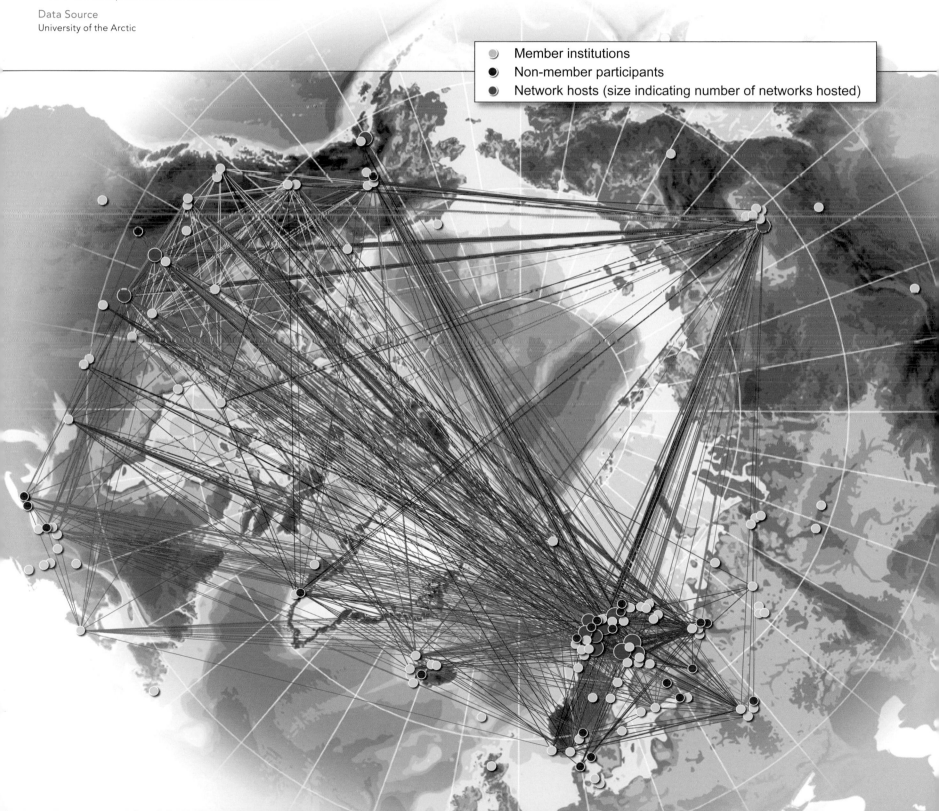

- Member institutions
- Non-member participants
- Network hosts (size indicating number of networks hosted)

Sea Level Rise Using Spatial Analyst

California Department of Transportation
Sacramento, California, USA
By Ron Abelon, GISP, ATP

Contact
Ron Abelon
ron_abelon@dot.ca.gov

Software
ArcGIS 10 for Desktop, ArcGIS Spatial Analyst

Data Sources
Esri World Imagery, National Oceanic and Atmospheric Administration, US Geological Survey, Caltrans

California Department of Transportation (Caltrans) was required to conduct a climate change infrastructure adaptation strategy. In December 2011, the Office of CADD and Engineering GIS Support in Sacramento developed various GIS models to satisfy requirements for delineating batched sea level rise (SLR) scenarios.

Surpassing grid mathematics that are ordinarily employed worldwide, Caltrans Engineering GIS also developed postprocessing models that eliminated hydraulically disconnected zones, removed miscellaneous polygons, and filled small voids. The process significantly reduced the number of generated vertices while improving the results. The final datasets had fewer irregularities, were simpler to interpret, and were less cumbersome to publish.

The Abelon Grid-Polygon Method converted data back and forth from raster to vector whenever the requirements exceeded the format capabilities. Since models were developed in-house, Caltrans was able to manipulate variables and customize processes to perform studies that included confidence-level tests, vulnerability identification, and tail water scenarios. If variable updates were required, the entire 1,000-mile coastline dataset could be regenerated in days rather than months. An example of the final SLR output can be seen on this map. The red polygons denote where the state-owned highways intersect the water delineation. Levels in the map have been altered from the official values to dramatize a hypothetical sea level rise scenario.

Courtesy of California Department of Transportation.

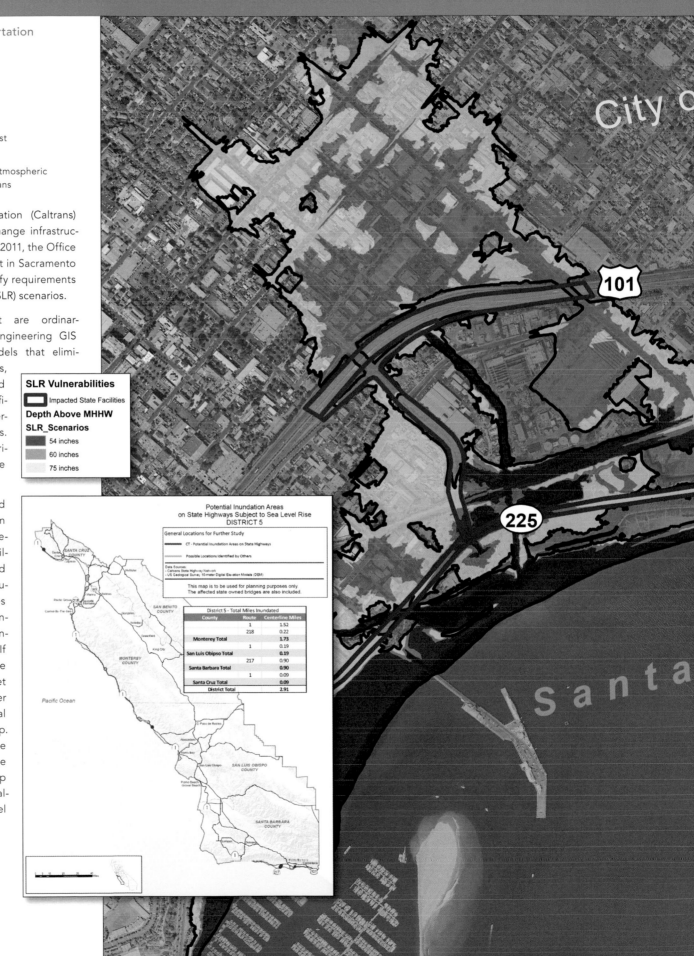

Mapping of the Ecological Network of the Regional Natural Park of the Volcanoes of Auvergne: Potential Ecological Network of Forest Environments

Asconit Consultants
Lyon, Rhône-Alpes, France
By Fabien Pezzato and Céline Thyriot

Contact
Rémy Martin
remy.martin@asconit.com

Software
ArcGIS Desktop 9.2, ArcGIS Spatial Analyst

Data Sources
PNR Volcans d'Auvergne, Asconit Consultants, IPAMAC, BD Topo IGN, BD Carthage

Since the 1992 Earth Summit, biodiversity has become an issue acknowledged by the whole international community. At the end of the 1990s, green and blue (land and water) corridor development projects started to emerge in France. Ecological corridors contribute to improving landscape quality and diversity, developing natural areas, and ensuring that they are taken into consideration in land planning

The Regional Natural Park of the Volcans d'Auvergne is composed of outstanding landscapes, fauna, and flora. As this territory does not really present ecological continuity problems, the two main issues were to maintain ecosystem quality and sustain ecological diversity. The aim of the assignment was to produce the map of the ecological networks, which led to the identification of potential nature hot spots of four types of subnetworks (meadow areas and moors, forest areas, aquatic ecosystems and wetlands, rock areas).

The GIS tools with spatial analysis methods have helped define the ecosystems' capacity to receive a maximum number of species and thus highlight continuity or break zones in view of the elaboration of an action strategy for their restoration or preservation. This map is the result of the modeling of forest area corridors. The analysis was based on the use of ecological data, technical know-how, and expertise. The areas of poor connectivity potential are not attractive and are far from hot spot zones.

Courtesy of Regional Natural Park of Volcans d'Auvergne.

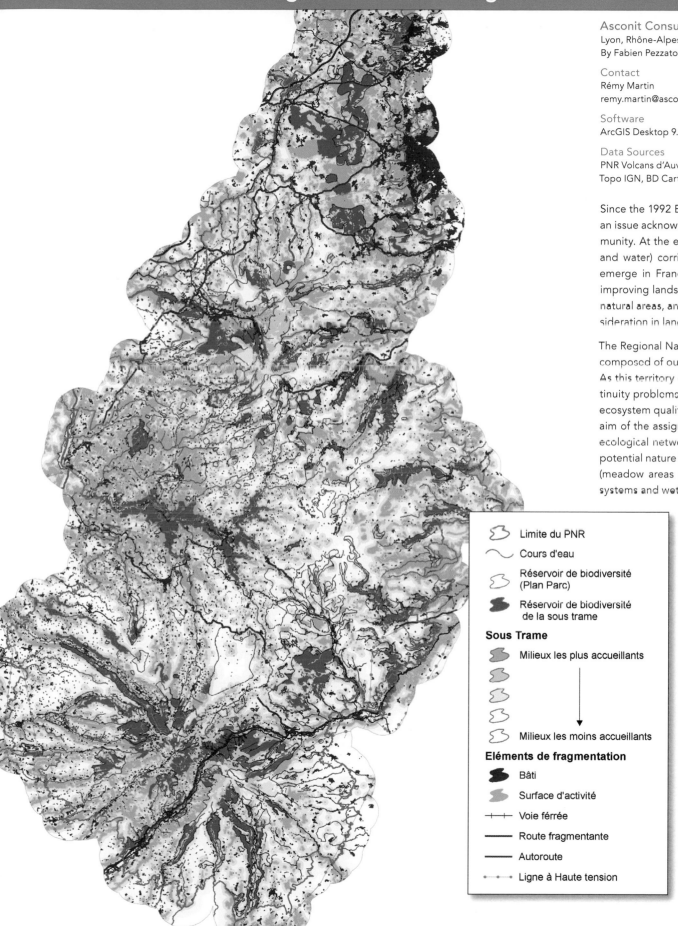

Legend

Limite du PNR
Cours d'eau
Réservoir de biodiversité (Plan Parc)
Réservoir de biodiversité de la sous trame

Sous Trame

Milieux les plus accueillants

Milieux les moins accueillants

Eléments de fragmentation

Bâti
Surface d'activité
Voie férrée
Route fragmentante
Autoroute
Ligne à Haute tension

Registration Mollusks in the Department of La Guajira, Colombia

Marine and Coastal Research Institute of Colombia
Santa Marta, Colombia
By Jose Eduardo Fuentes Delgado

Contact
Jose Eduardo Fuentes Delgado
mcfuentes2@hotmail.com

Software
ArcGIS Desktop 9.3.1

Data Sources
Marine and Coastal Research Institute of Colombia, Marine Biodiversity Information System in Colombia, A. Gracia

This map was created in the wildlife component framework for the Marine Coastal Atlas Department of La Guajira, Colombia, and uses available records within the Marine Biodiversity Information System. The map represents the spatial distribution of the families of mollusks that have been collected on several marine research projects in the area. This information aims to serve and facilitate the visualization and analysis for the identification of collection sites and potential distribution of marine malacofauna.

Mollusks are representatives of an important group of terrestrial, freshwater, and marine organisms, highly adapted to living in shallow and deep ocean and in distribution ranging from the tropics to the poles. According to existing records and available information, the estimated number of species found on the Colombian Guajira platform could comprise more than 900 taxa grouped into five classes.

Courtesy of Marine and Coastal Research Institute of Colombia.

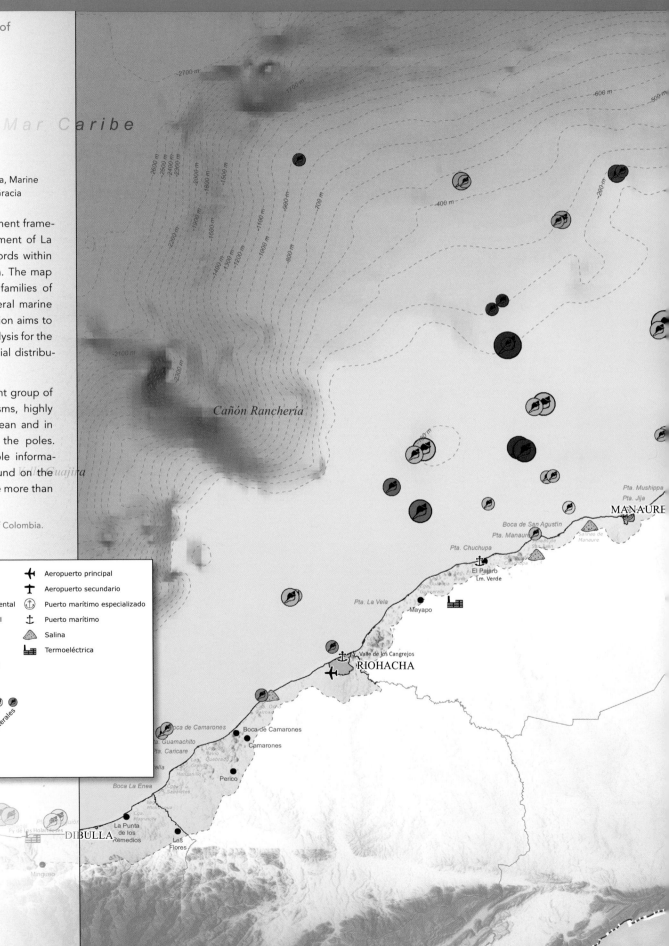

Cañón Macuira

Monjes Norte

Monjes Este
Monjes Sur

Pta. Gallinas
Pta. Taroa
Pta. Taroa
Pta. Huojojo
Lm. Huaripatu
Taroa
Pta. Huayapain
Pta. Shupehin
Pta. Parey
Pta. Aguja (Chitara)
Pto Chimare
Pta. Shuapia
Pta. Soldado
Pta. Cañon
Cb. Falso
Bh. Lepu
Bh. Honda
Pta. Ounorep
Pta. Jir...
Pta. Kayushpana
Puerto
Estrella
Bh. Honda
Pto Lodo
Salinas de Kemis
Pta. Chichibacoa
San José de
Bahía Honda
Salinas de
Agrasesh
Chichibacoa
Pta. de Media Luna
Salinas de
Malabe
Bocas de Apure
Pto Bolívar
Fria. Ripia
Bh. Portete
Pta. Ojo de Agua
Piñon de Azucar
Lm. La Mesa del Cabo
Cb. de la U
Salinas Kasimesh
Eda. Huaritche
Pto Portete
Cabo de la Vela
Serranía de Jarara
Pta. Kowshochom

Pta. Espada

Carrizal
Pta. Solipa

Pin De Cardón
Pta. Arenas
El Cardón
Dh. Tukuka
Puerto Lopez
Pta. Gorda

Pta. Castilletes
Castilotor
Bh. Cocinetas

URIBIA

Áreas marítimas pendientes
de delimitación

★

MAICAO

VENEZUELA

0   5   10   20
                        km
1:800.000

Spatial Distribution of Selected Tree Species in Europe

Techische Universität München (Technical University of Munich)
Munich, Germany
By Holm Seifert

Contact
Holm Seifert
Holm.M.Seifert@gmail.com

Software
ArcGIS for Desktop

Data Sources
NaturalEarthData; Bund für Naturschutz, Germany; European Environmental Agency; AFE Data: Botanical Museum of Helsinki; ISPRA Data: Bavarian regional office for forest and forest management ICP Forest Level 1, Italy; Meusel Data: Bavarian regional office

For the preservation and reestablishment of ecosystems, it is crucial to discover and analyze climate impacts and climate trends to ecosystems, especially to selected tree species. One indicator used to determine the impact of climate factors to areal distribution is trends in the abundance and distribution of selected tree species. Forest ecosystems as part of terrestrial ecosystems with their different areal tree distributions cover a wide range of Europe and play an essential role for future distribution shifts of tree species, triggered by climate factors. The distribution of the temperature and precipitation, as well as the underlying distribution of the current and potential tree species, are crucial factors.

This map shows the distribution of Scots pine (*Pinus sylvestris*), based on several input datasets for Europe. Light green areas describe actual distribution, where the tree actually occurs. Blue grid cells determine areas where the tree potentially should occur without any human influence but effectively couldn't be found in these areas. Orange grid cells symbolize areas where Scots pine actually occurs or potentially could occur and an intersection of actual and potential datasets take place. And finally, dark green areas visualize the distribution of the trees where all datasets—both actual distribution and potential distribution—match together. Besides the map, there are two graphics relating to the range area, the normalized area to total compliance and to quantity indexes.

The distribution datasets of *Pinus sylvestris* were now combined with the distributions of temperature and precipitation for Europe. The resulting map depicts the distribution of the species based on two climate factors, which were transformed together with the Tree Matrix datasets in a two-dimensional scatterplot matrix without spatial reference.

Courtesy of Techische Universität München.

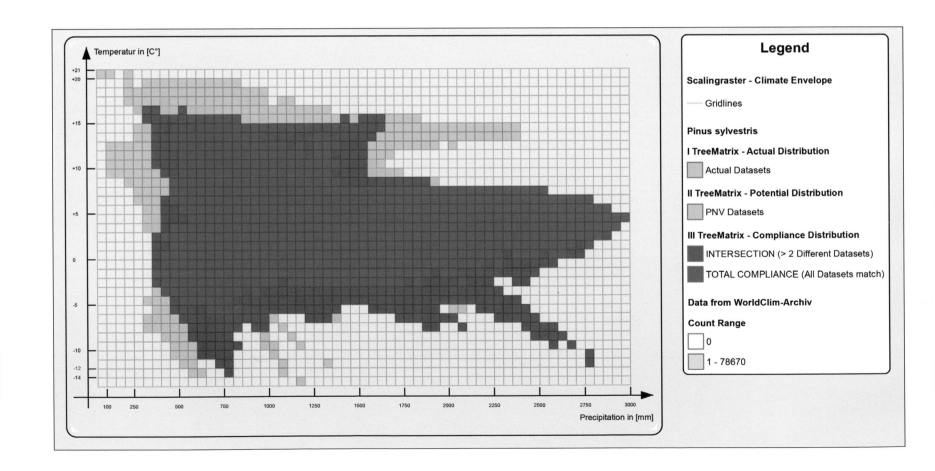

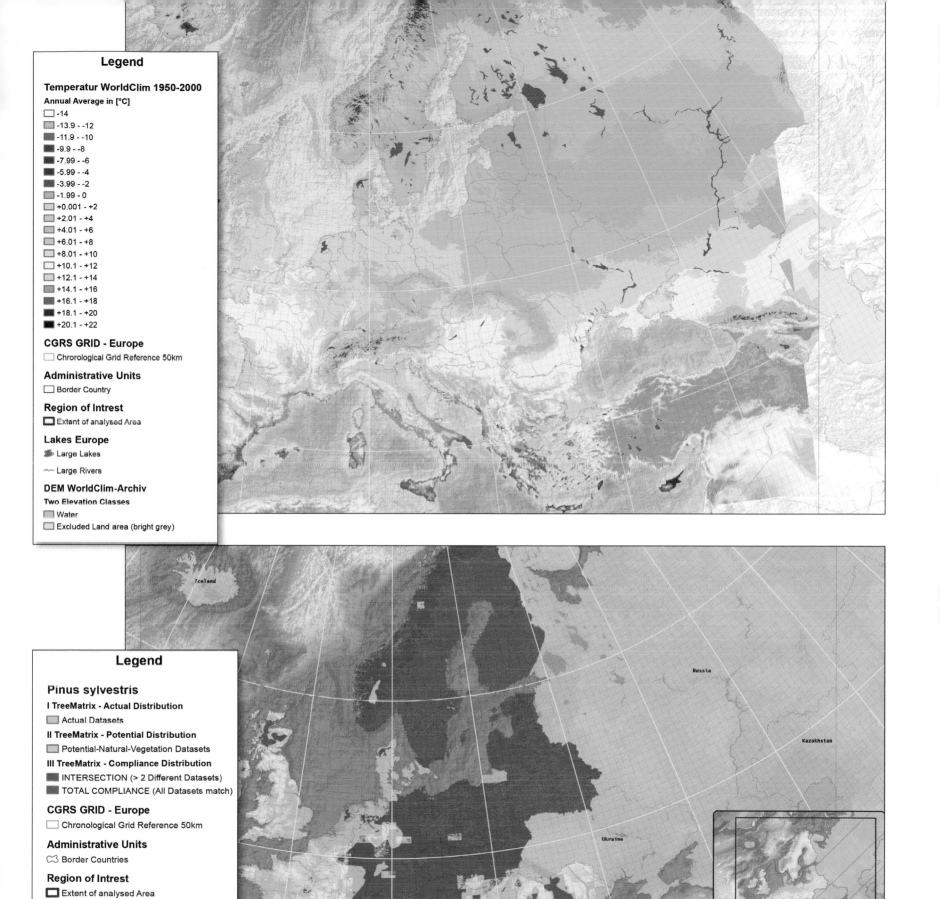

Legend

Temperatur WorldClim 1950-2000

Annual Average in [°C]
- -14
- -13.9 - -12
- -11.9 - -10
- -9.9 - -8
- -7.99 - -6
- -5.99 - -4
- -3.99 - -2
- -1.99 - 0
- +0.001 - +2
- +2.01 - +4
- +4.01 - +6
- +6.01 - +8
- +8.01 - +10
- +10.1 - +12
- +12.1 - +14
- +14.1 - +16
- +16.1 - +18
- +18.1 - +20
- +20.1 - +22

CGRS GRID - Europe
- Chrorological Grid Reference 50km

Administrative Units
- Border Country

Region of Intrest
- Extent of analysed Area

Lakes Europe
- Large Lakes
- Large Rivers

DEM WorldClim-Archiv

Two Elevation Classes
- Water
- Excluded Land area (bright grey)

Legend

Pinus sylvestris

I TreeMatrix - Actual Distribution
- Actual Datasets

II TreeMatrix - Potential Distribution
- Potential-Natural-Vegetation Datasets

III TreeMatrix - Compliance Distribution
- INTERSECTION (> 2 Different Datasets)
- TOTAL COMPLIANCE (All Datasets match)

CGRS GRID - Europe
- Chronological Grid Reference 50km

Administrative Units
- Border Countries

Region of Intrest
- Extent of analysed Area

Excluded Countries (Areas)
- Excluded Area by ISPRA Extent

Lakes Europe
- Large Lakes

Rivers Europe
- Large Rivers

Arctic Ocean Basemap

Nordpil
Stockholm, Sweden
By Hugo Ahlenius

Contact
Hugo Ahlenius
hugo.ahlenius@nordpil.com

Software
ArcGIS for Desktop, Global Mapper, Adobe Illustrator

Data Sources
CleanTOPO2, General Bathymetric Chart of the Oceans (GEBCO), International Bathymetric Chart of the Arctic Ocean (IBCAO), and Natural Earth (cross-blended hypsography for land areas)

This is a very detailed basemap for an interactive map service. The application focuses on geological data for the oceans in the greater circumpolar Arctic with the map serving as a backdrop. The design presents bathymetry and shaded relief for the oceans using the best available data sources and features muted colors so as not to conflict with other layers. This map was composited and rendered from various data sources that were merged into one dataset.

Courtesy of Hugo Ahlenius.

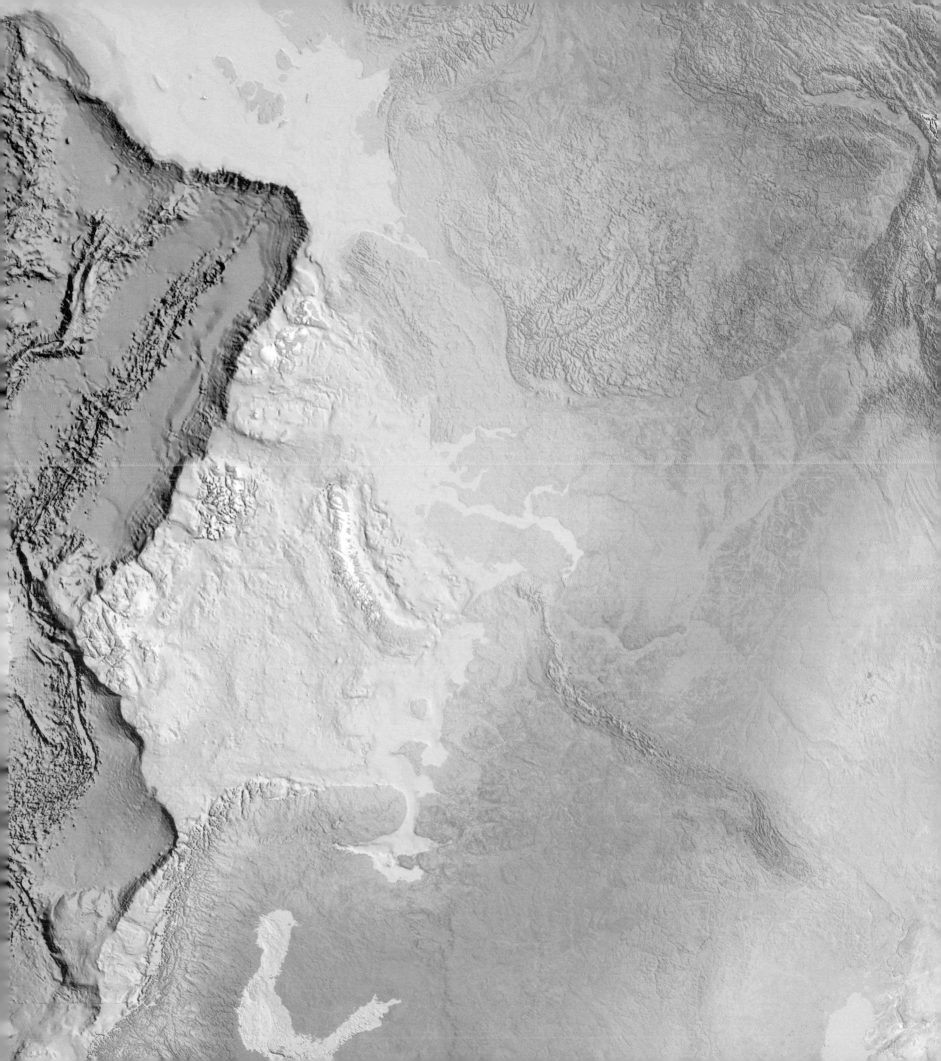

Cobb County Emergency 9-1-1 Call Analysis

Cobb County
Marietta, Georgia, USA
By Jennifer Lana, Brad Gordon, and Charles Fail

Contact
Jennifer Lana
jennifer.lana@cobbcounty.org

Software
ArcGIS for 10 Desktop, SQL 2008

Data Source
Cobb County

Cobb County, Georgia, estimates current costs, staff resources, and future budget allocations to maintain excellent public service operation levels for its police department. This sets a benchmark for understanding the link between designated Fiscal Impact Analysis Zones and the demand for public services in Cobb County, as well as for understanding spatial patterns in received response calls.

The overall spatial differences in received police response calls were evaluated for 2010 and 2012. The additional hot-spot analysis was based on call locations, identified regions within Cobb County where significant changes have occurred. This ongoing analysis enables management to make informed decisions on how to allocate resources and funding within defined regions of the county based on distribution patterns.

Courtesy of Cobb County, Georgia.

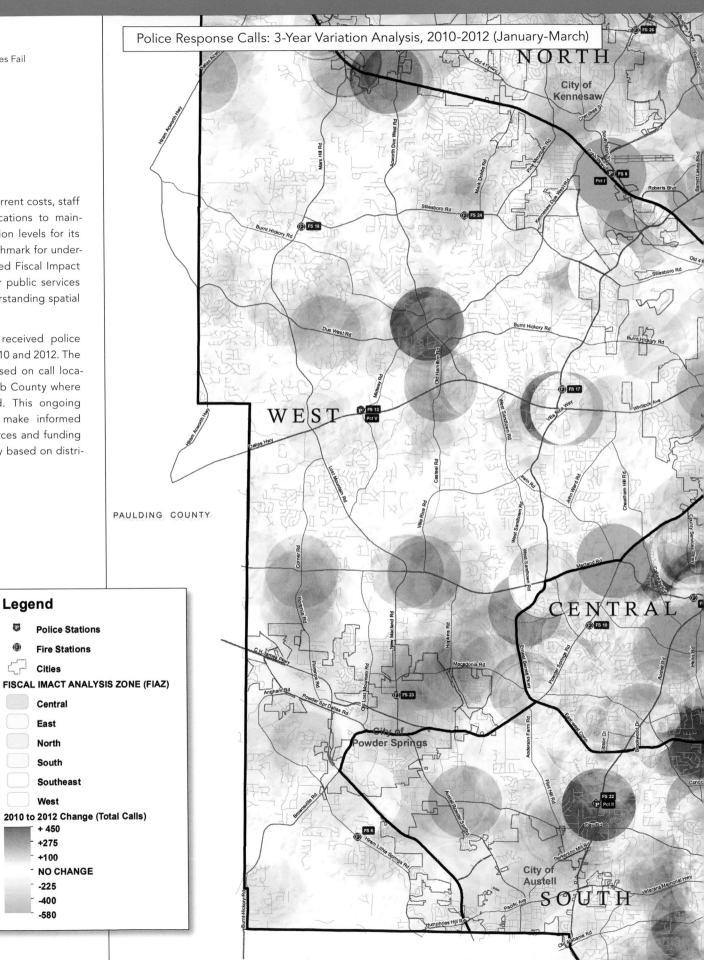

Police Response Calls: 3-Year Variation Analysis, 2010-2012 (January-March)

Legend

⬡ Police Stations

◉ Fire Stations

⬚ Cities

FISCAL IMACT ANALYSIS ZONE (FIAZ)

Central

East

North

South

Southeast

West

2010 to 2012 Change (Total Calls)

+ 450
+275
+100
NO CHANGE
-225
-400
-580

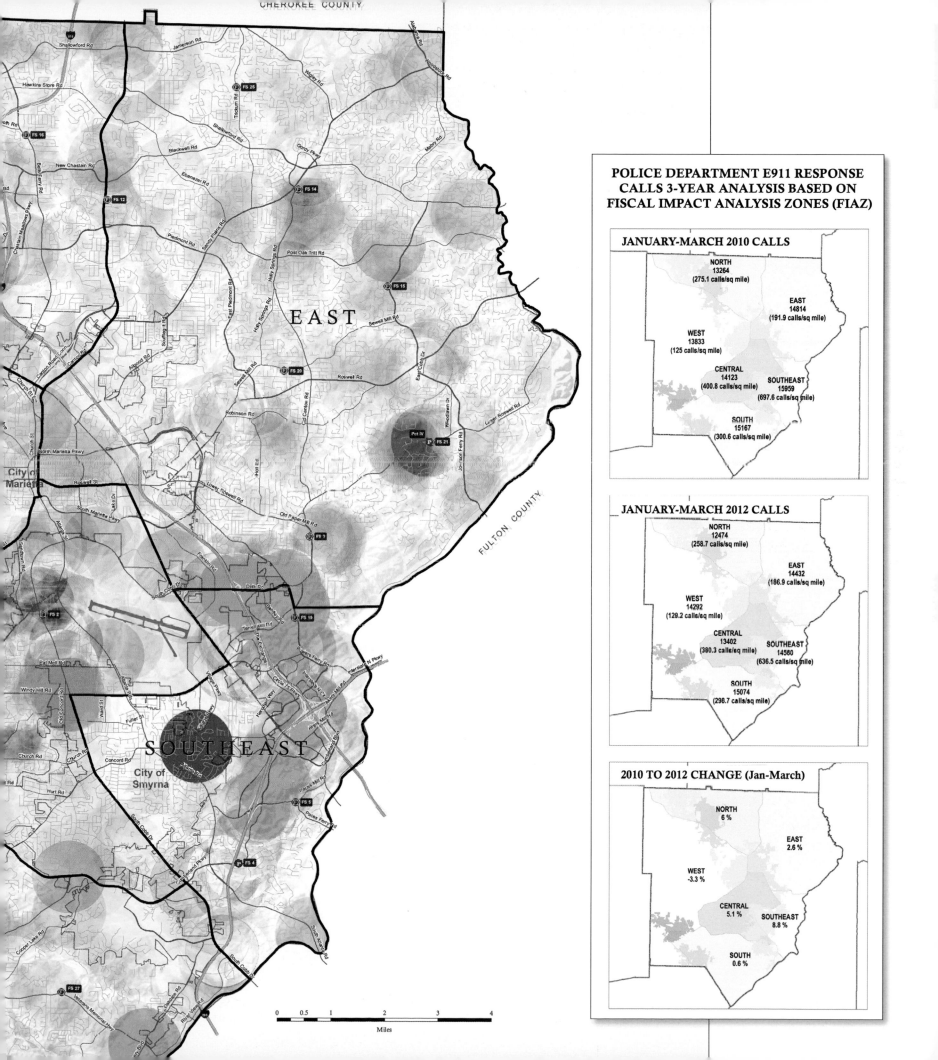

CHEROKEE COUNTY

FULTON COUNTY

EAST

City of Marietta

City of Smyrna

SOUTHEAST

0   0.5   1   2   3   4
Miles

POLICE DEPARTMENT E911 RESPONSE CALLS 3-YEAR ANALYSIS BASED ON FISCAL IMPACT ANALYSIS ZONES (FIAZ)

JANUARY-MARCH 2010 CALLS

NORTH
13264
(275.1 calls/sq mile)

EAST
14814
(191.9 calls/sq mile)

WEST
13833
(125 calls/sq mile)

CENTRAL
14123
(400.8 calls/sq mile)

SOUTHEAST
15959
(697.6 calls/sq mile)

SOUTH
15167
(300.6 calls/sq mile)

JANUARY-MARCH 2012 CALLS

NORTH
12474
(258.7 calls/sq mile)

EAST
14432
(186.9 calls/sq mile)

WEST
14292
(129.2 calls/sq mile)

CENTRAL
13402
(380.3 calls/sq mile)

SOUTHEAST
14560
(636.5 calls/sq mile)

SOUTH
15074
(298.7 calls/sq mile)

2010 TO 2012 CHANGE (Jan-March)

NORTH
6 %

EAST
2.6 %

WEST
-3.3 %

CENTRAL
5.1 %

SOUTHEAST
8.8 %

SOUTH
0.6 %

Medic Calls in Baltimore City

Baltimore City Fire Department
Baltimore, Maryland, USA
By Peter Hanna

Contact
Peter Hanna
peter.hanna@baltimorecity.gov

Software
ArcGIS 10 for Desktop

Data Source
Baltimore City, 2011

The Baltimore City Fire Department handles approximately 160,000 calls per year. Of those calls, 130,000 are medical aid calls. Asset management has become a major concern for the Emergency Medical Services (EMS) division. There have been times when the call volume was so high, there were no available EMS units to take calls and fire apparatus needed to stand by for a medic.

Many issues affect the availability of EMS units. Some of those issues are call volume and hospital wait times. By using the tools located in the Spatial Statistics toolbox, the calls are now more than just dots on a map. Calls have been broken down by different call types, such as shootings, cuttings, and cardiac arrests, to show where the majority of the city's calls are occurring. The calls have also been grouped by hour of day to see how far units are responding from their stations. The data has shown that the busy area of the city is around the downtown area. Units are being pulled into this area to help with volume, which, in turn, affects response times to other areas of Baltimore.

More analysis is needed, but the tools located in the Spatial Statistics toolbox have helped the Baltimore Fire Department better understand its mission, manage its assets, and determine how to better serve the citizens of Baltimore.

Courtesy of Peter Hanna.

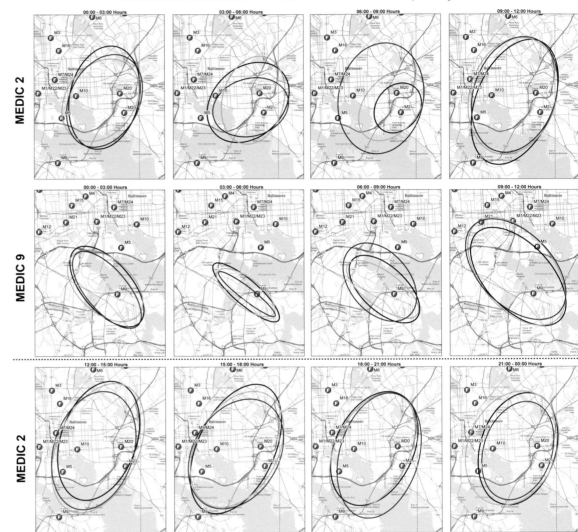

FY 2011 Medic 2 and Medic 9 - 1 Standard Deviational Ellipses by Hours

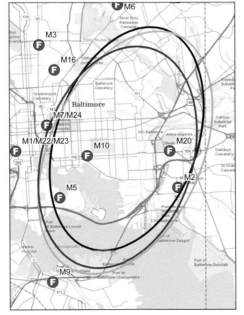
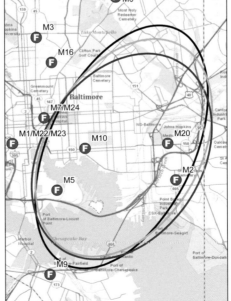
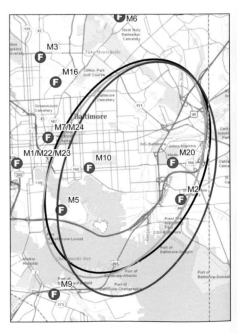

Hurricane Irene Response

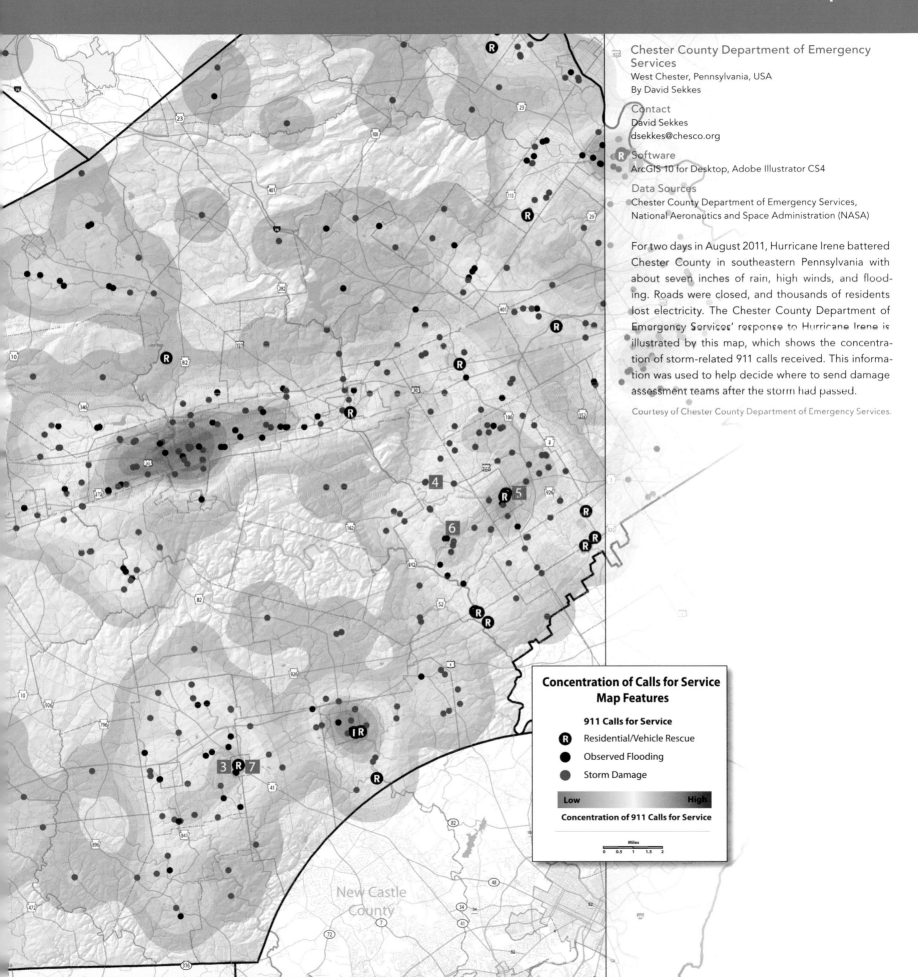

Chester County Department of Emergency Services
West Chester, Pennsylvania, USA
By David Sekkes

Contact
David Sekkes
dsekkes@chesco.org

Software
ArcGIS 10 for Desktop, Adobe Illustrator CS4

Data Sources
Chester County Department of Emergency Services,
National Aeronautics and Space Administration (NASA)

For two days in August 2011, Hurricane Irene battered Chester County in southeastern Pennsylvania with about seven inches of rain, high winds, and flooding. Roads were closed, and thousands of residents lost electricity. The Chester County Department of Emergency Services' response to Hurricane Irene is illustrated by this map, which shows the concentration of storm-related 911 calls received. This information was used to help decide where to send damage assessment teams after the storm had passed.

Courtesy of Chester County Department of Emergency Services.

Concentration of Calls for Service Map Features

911 Calls for Service

Ⓡ Residential/Vehicle Rescue

● Observed Flooding

● Storm Damage

Low   High

Concentration of 911 Calls for Service

Miles
0   0.5   1   1.5   2

Integrating Multiple Hazard Assessments to More Effectively Convey Potential Impacts of Landslides in El Salvador

University of Alabama in Huntsville
Huntsville, Alabama, USA
By Eric R. Anderson

Contact
Eric Anderson
eric.anderson@nsstc.uah.edu

Software
ArcGIS 10 for Desktop, LAHARZ for ArcInfo Workstation

Data Sources
US Geological Survey (USGS), NASA, USAID, CATHALAC, SERVIR, BGR (Germany), CEPREDENAC (Central America), INETER (Nicaragua), SNET (El Salvador), INSIVUMEH & CONRED (Guatemala), UNITEC & COPECO (Honduras)

Every year, landslides claim lives, property, and infrastructure in El Salvador. Landslides present a diverse set of hazards, and there are studies that address each kind. Two prominent studies focus on (1) identifying the potential trigger points of slope failures and (2) delineating landslide inundation zones. Ironically, since these two studies depict different characteristics of mass movements, they can convey incongruous or even conflicting information to development and disaster management officials. Trigger analyses only identify where slope failures may initiate but do not show areas susceptible to debris inundation. Inundation zone maps show potentially affected areas but do not consider trigger points. In El Salvador, inundation zone maps only exist around volcanoes but do not cover the extensive mountainous regions.

The solutions proposed here show how to (1) link the trigger maps with current inundation zone maps to have a better understanding of which areas are more exposed to real landslide hazards and (2) identify additional unknown inundation zones for the many potential shallow landslides throughout the country. This map represents cooperative efforts among the University of Alabama in Huntsville's Earth System Science program; the National Aeronautics and Space Administration (NASA); the US Agency for International Development (USAID); and international partners such as the Water Center for the Humid Tropics of Latin America and the Caribbean (CATHALAC) and the Ministry of the Environment and Natural Resources (MARN) of El Salvador, through the Regional Visualization and Monitoring System (SERVIR).

Courtesy of University of Alabama in Huntsville, NASA/USAID SERVIR.

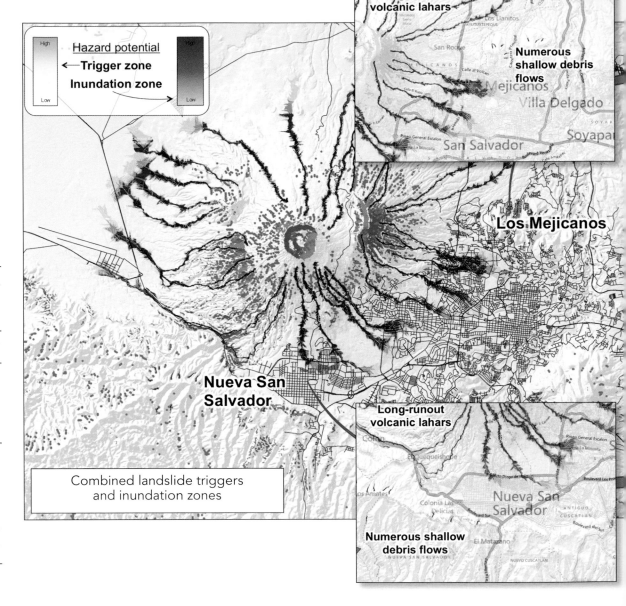

Hazard potential
← Trigger zone
Inundation zone

Long-runout volcanic lahars

Numerous shallow debris flows

Los Mejicanos

Nueva San Salvador

Long-runout volcanic lahars

Numerous shallow debris flows

Combined landslide triggers and inundation zones

Landslide hazards
in El Salvador

Landslide hazard potential
Low   Medium   High   Very High

Landslide hazards
in San Salvador Metropolitan Area

The vast majority of San Salvador's development appears to be near but not actually located in areas of very high landslide hazard potential.

On the other hand, potential debris flows appear to pose a serious threat to many of the capital city's populace. In fact, past landslides have inundated these areas.

Landslide trigger zones
"Where will they come from?"

Debris inundation zones
"Where will they go?"

Landslides and Floods in El Salvador, November 2009

Before and After Analysis

Formosat-2

False color

GUADALUPE

VERAPAZ

TEPETITAN

February 9, 2008

GUADALUPE

VERAPAZ

TEPETITAN

November 11, 2009

University of Alabama in Huntsville
Huntsville, Alabama, USA
By Africa I. Flores Cordova

Contact
Africa I. Flores Cordova
africa.flores@nsstc.uah.edu

Software
ArcGIS 10 for Desktop, SERVIR Viz

Data Sources
Water Center for the Humid Tropics of Latin America and the Caribbean (CATHALAC), Formosat image © 2009. Dr. Cheng-Chien Liu, National Cheng-Kung University, and Dr. An-Ming Wu, National Space Organization, Taiwan; Directorate-General of the Environmental Observatory (DGOA), Ministry of Environment and Natural Resources of El Salvador

This map depicts the areas affected by landslides and floods due to heavy rains during November 2009 in El Salvador. The soil saturation due to the constant rains, together with the steep terrain, unleashed mud slides and lahars from San Vicente Volcano, sending boulders and other debris downstream and causing rivers to burst their banks. This information has been used by the Salvadoran government to redraw zoning areas and direct reconstruction efforts.

Courtesy of University of Alabama in Huntsville, NASA/USAID SERVIR

Tepetitan

2,000 m

⬜	**Landslides**	**42 Ha.**
⬛	**Lahar inundation zone**	**258 Ha.**

Estimation of the affected areas within the extension of the satellite image shown above.

Tropical Storm and Hurricane Strikes per Municipality and Department in Central America, 1851–2009

University of Alabama in Huntsville
Huntsville, Alabama, USA
By Eric R. Anderson

Contact
Eric Anderson
eric.anderson@nsstc.uah.edu

Software
ArcGIS 10 for Desktop

Data Sources
International Best Track Archive for Climate Stewardship,
US Agency for International Development, Water Center
for the Humid Tropics of Latin America and the Caribbean,
Regional Visualization and Monitoring System, European
Space Agency GlobCover, National Oceanic and
Atmospheric Administration National Hurricane Center,
National Aeronautics and Space Administration Blue Marble

Due to the population's underlying vulnerability and its direct exposure to tropical weather events, Central America is one of the regions most affected by climate-related hazards. This map shows the number of direct hurricane and tropical storm strikes per municipality in Central America, giving citizens and government officials a clear picture of the places most frequently hit over the past 160 years.

Tropical cyclone center points from 1851 to 2009 were obtained from the International Best Track Archive for Climate Stewardship (IBTrACS 2010) and prepared using the methodology developed by Jarrell, Hebert, and Mayfield (1992; updated in the National Hurricane Center [NHC] 2010). Administrative limits were provided by the Water Center for the Humid Tropics of Latin America and the Caribbean through the Regional Visualization and Monitoring System (SERVIR), which had obtained the national-level information from environmental ministries. Along with ArcGIS analysis and data management tools, this composition relied on the free ET GeoWizards and Hawth's Tools extensions.

This map was created in a graduate-level GIS and remote-sensing course in the University of Alabama in Huntsville's Earth System Science program. It has also been incorporated into the SERVIR web portal, which reflects the collaborative efforts of the National Aeronautics and Space Administration (NASA), the US Agency for International Development, and international partners to promote the use of earth observation data for better environmental management and decision making.

Courtesy of University of Alabama in Huntsville, NASA/USAID SERVIR.

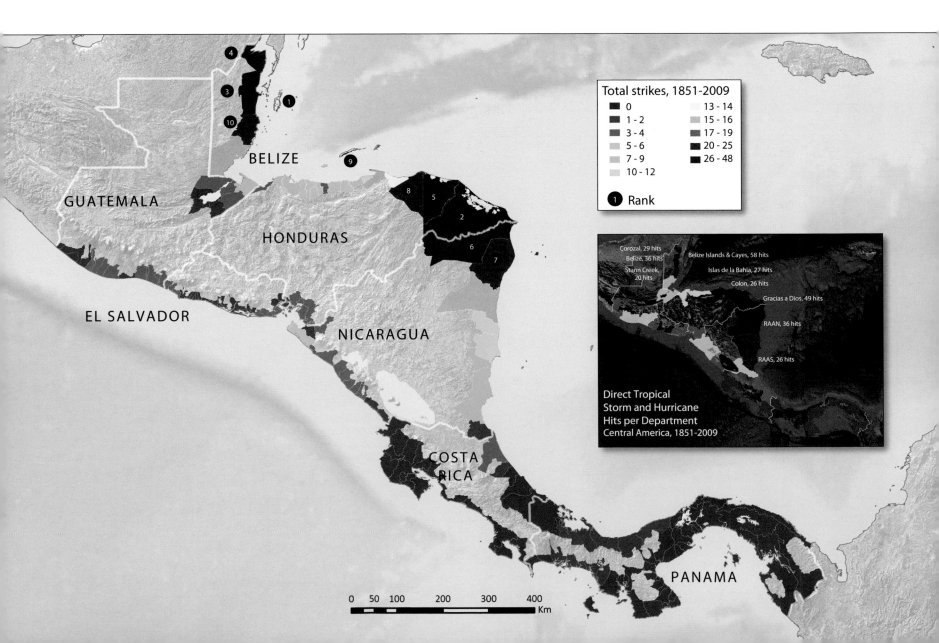

Total strikes, 1851-2009

0	13 - 14
1 - 2	15 - 16
3 - 4	17 - 19
5 - 6	20 - 25
7 - 9	26 - 48
10 - 12	

1 Rank

Direct Tropical Storm and Hurricane Hits per Department Central America, 1851-2009

Corozal, 29 hits
Belize, 36 hits
Stann Creek, 20 hits
Belize Islands & Cayes, 58 hits
Islas de la Bahia, 27 hits
Colon, 26 hits
Gracias a Dios, 49 hits
RAAN, 36 hits
RAAS, 26 hits

GUATEMALA
BELIZE
HONDURAS
EL SALVADOR
NICARAGUA
COSTA RICA
PANAMA

0 50 100   200   300   400
Km

Seoul National University
Seoul, South Korea
By Jin Son, Jangwon Suh, Sangho Lee, and
Hyeong-Dong Park

Contact
Hyeong-Dong Park
giscity@hanmail.net

Software
ArcGIS Desktop 9.3

Data Source
National Spatial Information Clearinghouse of South Korea

This map shows landslide susceptibility indexes in Seoul, South Korea, based on the frequency ratio of topographic and environmental factors of preexisting landslide sites. The target area is Mount Umyeon and the Seoul area, where there had been a considerable amount of property damage and eighteen casualties from landslides caused by local heavy rainfall in June 2011. Frequency ratios could be calculated with topographic factors mapped with Advanced Spaceborne Thermal Emission and Reflection Radiometer Global Digital Elevation Model as well as soil, forest, and land-use maps acquired from the Water Management Information System in South Korea. This map was obtained by analyzing correlations between frequency ratio and maximum hourly rainfall data from the Automated Weather System. Priorities of risk reduction activities for further landslide hazards were then established.

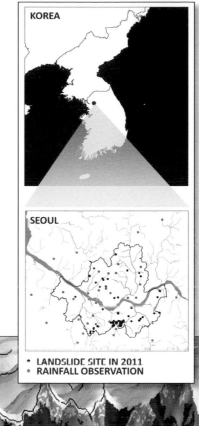

% OF LANDSLIDE SUSCEPTIBILITY INDEX

Top	80 – 100 Percent	
Top	60 – 80 Percent	
Top	40 – 60 Percent	
Top	20 – 40 Percent	
Top	0 – 20 Percent	

LANDSLIDE SITE IN 2011

Seasonal Tornado Density

Federal Emergency Management Agency
Region V
Chicago, Illinois, USA
By Thomas Griffin

Contact
Jesse Rozelle
jesse.rozelle@fema.dhs.gov

Software
ArcGIS 10 for Desktop

Data Sources
Esri, NOAA, FEMA

This seasonal tornado density map displays an analysis for Federal Emergency Management Agency (FEMA) Region V (Illinois, Indiana, Michigan, Minnesota, Ohio, and Wisconsin). Data was taken from the National Oceanic and Atmospheric Administration (NOAA) website about every recorded tornado since 1951. The track for every tornado that entered a Region V state was collected, and a line density was created. The data for these tornadoes also included information on when each tornado occurred, allowing the tornadoes to be further analyzed based on the season in which they occurred.

The purpose of this map was to identify any possible trends, take a closer look at relevant data, and tell a story through the use of GIS. Though tornadoes are very difficult to predict with any degree of accuracy, interesting assumptions can be deduced quickly and easily by looking at the data when it is visually represented. For instance, the map shows that tornadoes occurring during the winter months tend to be farther south where the weather is warmer, while tornadoes occurring during the summer months tend to spread farther north. The ability to see and understand this type of information better using GIS is potentially lifesaving and is being used within FEMA more and more every day.

Courtesy of Federal Emergency Management Agency Region V.

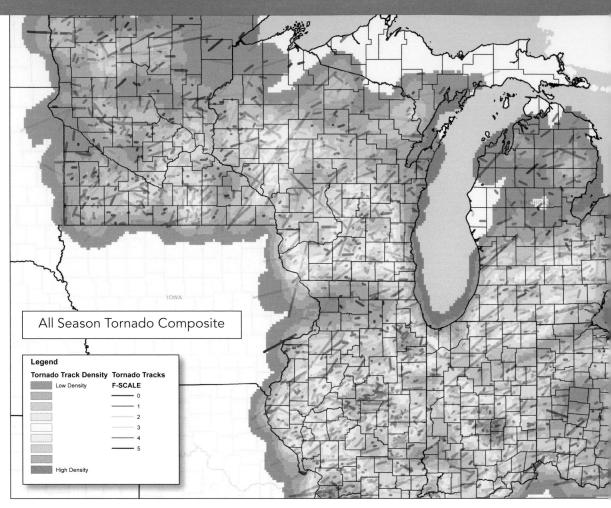

All Season Tornado Composite

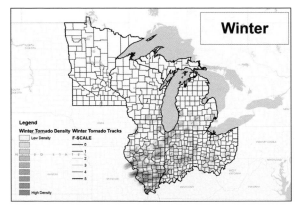

Winter

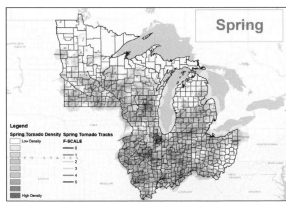

Spring

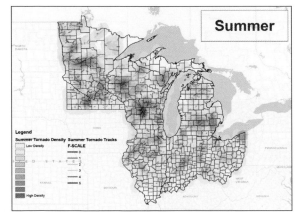

Summer

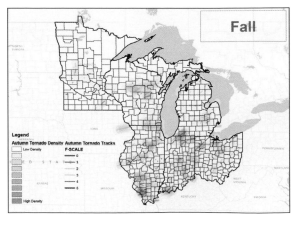

Fall

Joplin Tornado Damage Assessment

Federal Emergency Management Agency Region VII
Kansas City, Missouri, USA
By Derek Duskin

Contact
Jesse Rozelle
jesse.rozelle@fema.dhs.gov

Software
ArcGIS 10 for Desktop

Data Source
City of Joplin, City of Duquesne, Highway Safety Improvement Program, NLT/ImageCat, Jasper County, Missouri Department of Transportation, NOAA, NWS, State Emergency Management Agency, Surdex, US Army Corps of Engineers, National Geospatial-Intelligence Agency

The Joplin Tornado analysis map displays the extent and magnitude of the disaster, along with how remote-sensing products became an invaluable tool for both response and recovery. An assessment by New Light Technologies (NLT)/ImageCat, Inc., identified over 7,500 buildings damaged by the tornado. The building damage assessment displayed in the legend was created using a weighted kernel density found within ArcGIS Spatial Analyst tools with ImageCat point data that classified structures from damage classification (limited, moderate, extensive, and catastrophic).

Furthermore, the map displays the correlation of the Enhanced Fujita scale and the ImageCat data. The map displays the Expedited Debris Removal Zone in which ImageCat data was also a valuable tool for the Federal Emergency Management Agency (FEMA) and the City of Joplin in carving the cost share boundary. The Joplin Tornado Path, reports, and Enhanced Fujita scale were supplied by the National Weather Service (NWS) and the National Oceanic and Atmospheric Administration (NOAA). The before and after imagery was supplied by NOAA and Surdex.

Courtesy of Federal Emergency Management Agency Region VII.

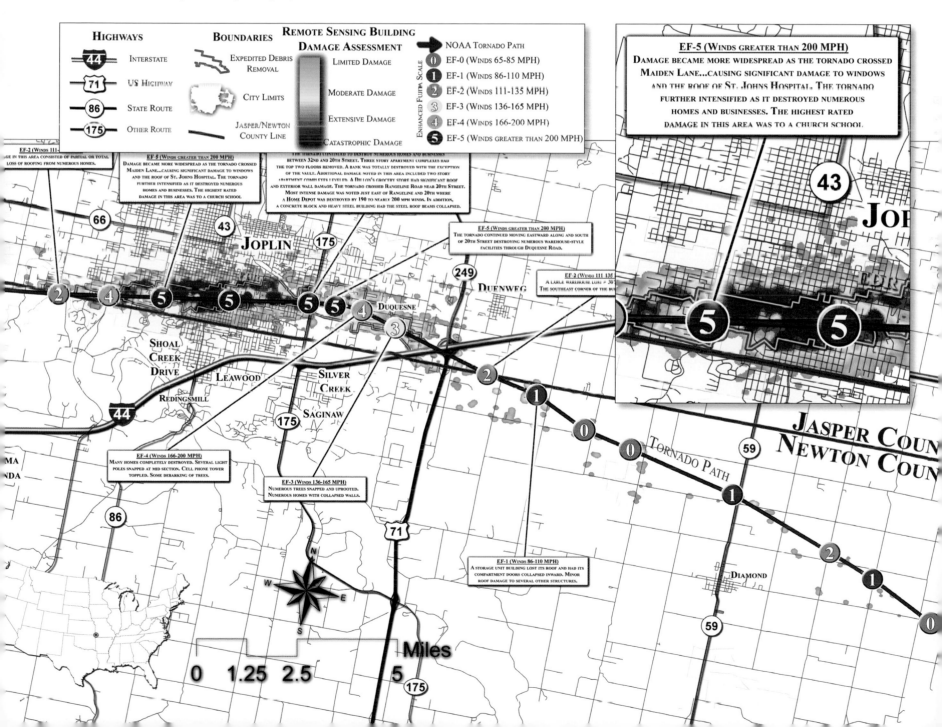

Hazus Saudi Arabia Earthquake Loss Estimation Study for the Tabuk Region

Federal Emergency Management Agency
Region VIII
Denver, Colorado, USA
By Jesse Rozelle and Doug Bausch (FEMA Region VIII) and
Hani M. Zahran (Saudi Geological Survey)

Contact
Jesse Rozelle
jesse.rozelle@fema.dhs.gov

Software
ArcGIS 10 for Desktop, Hazus

Data Sources
NASA Shuttle Radar Topography Mission, FEMA, FEMA
Hazus earthquake model, Saudi Geological Survey

The Federal Emergency Management Agency (FEMA) and Saudi Geological Survey completed a joint collaboration to apply FEMA's Hazus earthquake loss model to the Tabuk region of Saudi Arabia. This included replacing the US data with data from the Tabuk region, including information on buildings in Tabuk City and Haql as well as updated geologic information. The Hazus model provides loss modeling for earthquakes, floods, and hurricanes, and the results are used by emergency managers and decision makers to better plan for and reduce the impacts of future disasters.

The assessment for the Tabuk region included evaluating the potential impacts of a magnitude 7.3 Gulf of Aqaba earthquake near Haql and magnitude 6.0 and 6.5 scenarios located along a fault southwest of Tabuk City. Each of the scenarios is in an area of historic earthquake epicenters. The results indicate potential building economic losses of 1 billion Saudi Arabian riyals and potential deaths ranging from twenty to forty-five, mostly in the Haql area from a Gulf of Aqaba magnitude 7.3 earthquake. The fault scenarios nearer Tabuk City indicate a potential loss range from 1.6 to 7.1 billion Saudi Arabian riyals and potential deaths ranging from 20 to 600.

Courtesy of Federal Emergency Management Agency Region VIII and Saudi Geological Survey.

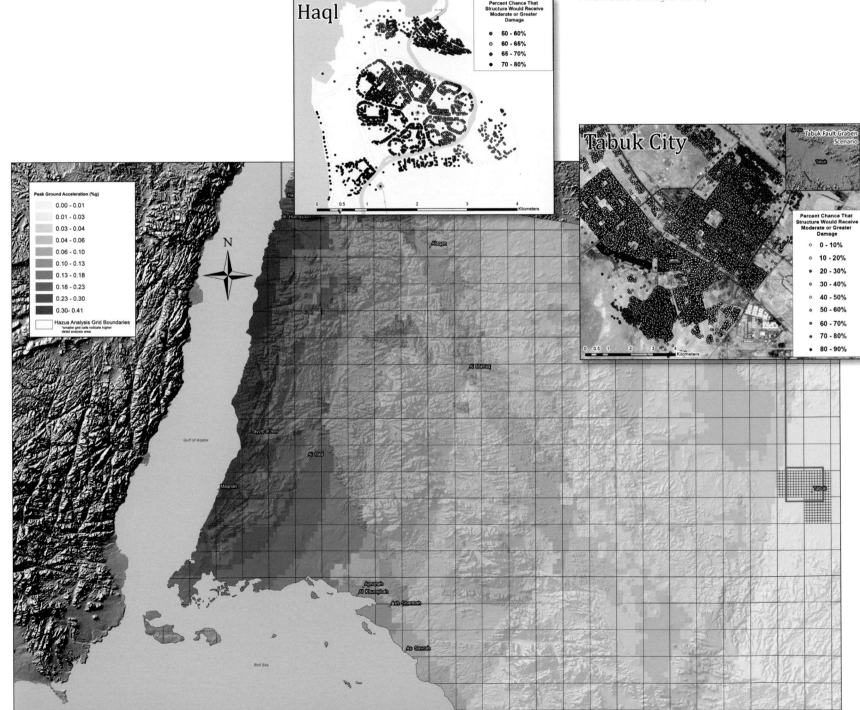

Evaluation of the Volcanic Hazard on the Piton de la Fournaise Volcano (La Réunion Island)

A

B

C

Laboratoire d Informatique de Grenoble (Grenoble Informatics Laboratory)
Saint Martin d Heres, Rhône-Alpes, France
By Paule Annick Davoine, Cécile Saint Marc, Andrea Di Muro, Thomas Staudacher, Patrice Boissier, Laurent Michon, and Nicolas Villeneuve

Contact
Paule Annick Davoine
Paule-Annick.Davoine@imag.fr

Software
ArcGIS for Desktop, ArcGIS Spatial Analyst

Data Sources
Observatoire Volcanologique du Piton de la Fournaise, IGN France

The Piton de la Fournaise is a Hawaiian effusive volcano located on La Réunion Island in the Indian Ocean. With seventy-five eruptions since 1972, it is considered one of the most active volcanoes in the world. Eruptions mostly occur in the Enclos Fouqué caldera, which prevents access to the top of the volcano for tourists. Furthermore, some eruptions regularly happen outside the caldera (1977, 1986, 1998, 2007) and threaten urban areas and road infrastructure.

The Volcanologic Observatory of the Piton de la Fournaise (OVPF), which belongs to the Institut de Physique du Globe de Paris (IPGP), keeps watch on the volcano and studies its functioning. As part of a study to evaluate volcanic hazard, the OVPF has partnered with the Laboratoire d Informatique de Grenoble to design and produce a GIS to map past phenomenon that have occurred during the eruptions: lava flows, cones, eruptive and noneruptive faults, and tephras.

GIS has spatialized the eruptive mechanism and mapped the hazards associated with each phenomenon. Among these, lava flow hazard is the most representative of volcanic hazard. Its characterization is based on a map of lava flows (a) and on the identification of the most frequently impacted areas (c). Nevertheless, the temporal aspect of this geographic data complicates cartographic representations. As a consequence, some semiologic reflection is needed to map chrono-spatial data (b). Then, the combination of the phenomenological hazards will create a map of global volcanic hazards.

Courtesy of Laboratoire d Informatique de Grenoble STEAMER group, Volcanologic Observatory of the Piton de la Fournaise, Institut de Physique du Globe de Paris.

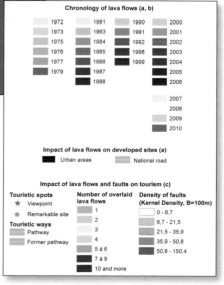

Chronology of lava flows (a, b)

1972	1981	1990	2000
1973	1983	1991	2001
1975	1984	1992	2002
1976	1985	1998	2003
1977	1986	1999	2004
1979	1987		2005
1988			2006

2007
2008
2009
2010

Impact of lava flows on developed sites (a)
Urban areas   National road

Impact of lava flows and faults on tourism (c)

Touristic spots
★ Viewpoint
✳ Remarkable site

Touristic ways
Pathway
Former pathway

Number of overlaid lava flows
1
2
3
4
5 à 6
7 à 9
10 and more

Density of faults (Kernel Density, B=100m)
0 - 6,7
6,7 - 21,5
21,5 - 35,9
35,9 - 50,8
50,8 - 150,4

Minot, North Dakota: Souris (Mouse) River Extent, Depth of Water at Structures, and Estimated Losses—Post Event Imagery

Federal Emergency Management Agency
Region VIII
Denver, Colorado, USA
By Jesse Rozelle

Contact
Jesse Rozelle
jesse.rozelle@fema.dhs.gov

Software
ArcGIS 10 for Desktop, Hazus

Data Sources
North Dakota Department of Emergency Services, US Army Corps of Engineers, USGS, New Light Technologies/ImageCat, Houston Engineering, Pictometry, City of Minot, FEMA, FEMA Hazus flood model, 2011 GeoEye NextView License

The City of Minot, North Dakota, population 40,888 (2010), is located along the Souris (Mouse) River in Ward County, roughly 50 miles south of the Canadian border. In June 2011, abnormally high amounts of rainfall, a record snowpack, saturated soils, and reservoir capacities being met led to record river flows along the Mouse River, impacting over 4,000 structures in Minot. The Federal Emergency Management Agency (FEMA) Region VIII Mitigation GIS, working with several other partners, conducted a detailed site specific analysis of the structures impacted.

Parcel data, lidar-generated depth grids, US Geological Survey (USGS) high water marks, and oblique and satellite imagery were all used in conjunction with depth damage functions provided by FEMA's Hazus flood model to estimate detailed flood losses to the structures impacted along the Mouse River in Minot. The full analysis was completed less than a week after the crest of the flood event, greatly assisting in response and recovery efforts in Minot.

Courtesy of Federal Emergency Management Agency Region VIII.

Map Legend

High Water Marks (NAVD '88)

Minot Evacuation Zones

US National Grid

Confirmed Protected Areas (FEMA/USGS Field Survey)

Structures Affected - 4,165

*Estimated number of structures affected in the Minot, ND area based on modeling and imagery analysis

Current Effective FIRM

*Digitized from current Flood Insurance Rate Map (FIRM), dated January 19, 2000.

SFHA - 100 Year Floodplain
Affected Structures - 35

500 Year Floodplain
Affected Structures - 4,899

Water Depth at Structure

- 0 - 1 ft
- 1 - 4ft
- 4 - 6 ft
- 6 - 7ft
- 7 - 10ft
- > 10ft

0.25   0.5   0.75   1
Miles

Impervious Surface Analysis

City of Houston

Houston, Texas, USA
By Larry Nierth, GISP

Contact
Larry Nierth
larry.nierth@houstonTX.gov

Software
ArcGIS for Desktop

Data Sources
Harris County Flood Control District, City of Houston
Enterprise Geodatabase

This map shows Houston's citywide impervious surfaces and illustrates the surface change from 2010 to 2012. Areas around town that had their surfaces change from impervious to pervious represented increased demolition or promotion of open space. By contrast, areas that had their surfaces type change from pervious to impervious represented locations of buildup or construction. Analyzing these trends is important because these changes have significant impact on the city's drainage system.

The impervious surface rasters were generated as 1-foot-per-pixel data from 1-foot color infrared (CIR) orthoimagery. Afterward, the rasters were reclassified and mosaicked in ArcGIS, and building footprints

were added to the impervious classification. A zonal analysis was run against all citywide parcels to get total square footage of pervious and impervious area per parcel.

The data paved the way to a pay-as-you-go approach to maintaining Houston's vast drainage system. This data is the foundation for the collection of over $125 million in annual revenues over the program's life cycle. Everyone inside the city pays a fair share into the drainage fund. The actual fees incurred are based on whether the property is residential, whether the property is serviced by a curb and gutter or an open ditch drainage system, and the total square footage of impervious surface area in the parcel.

Courtesy of Larry Nierth.

A
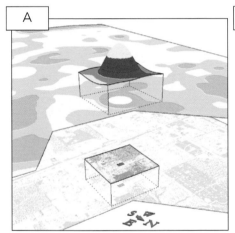

B
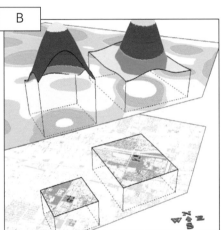

C
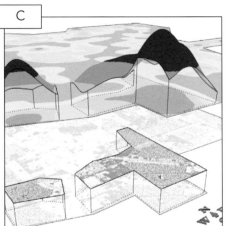

D
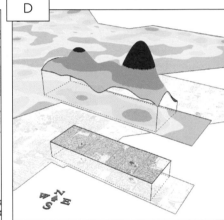

Density Shift From Impervious to Pervious Surface Type
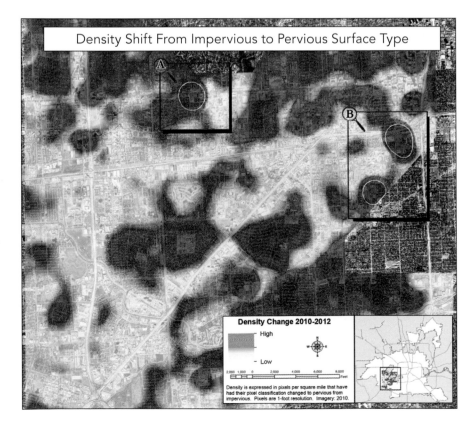

Density Shift From Pervious to Impervious Surface Type
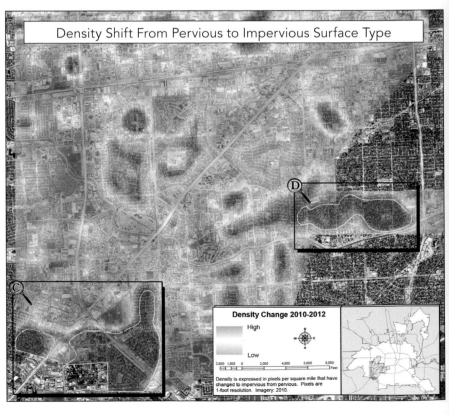

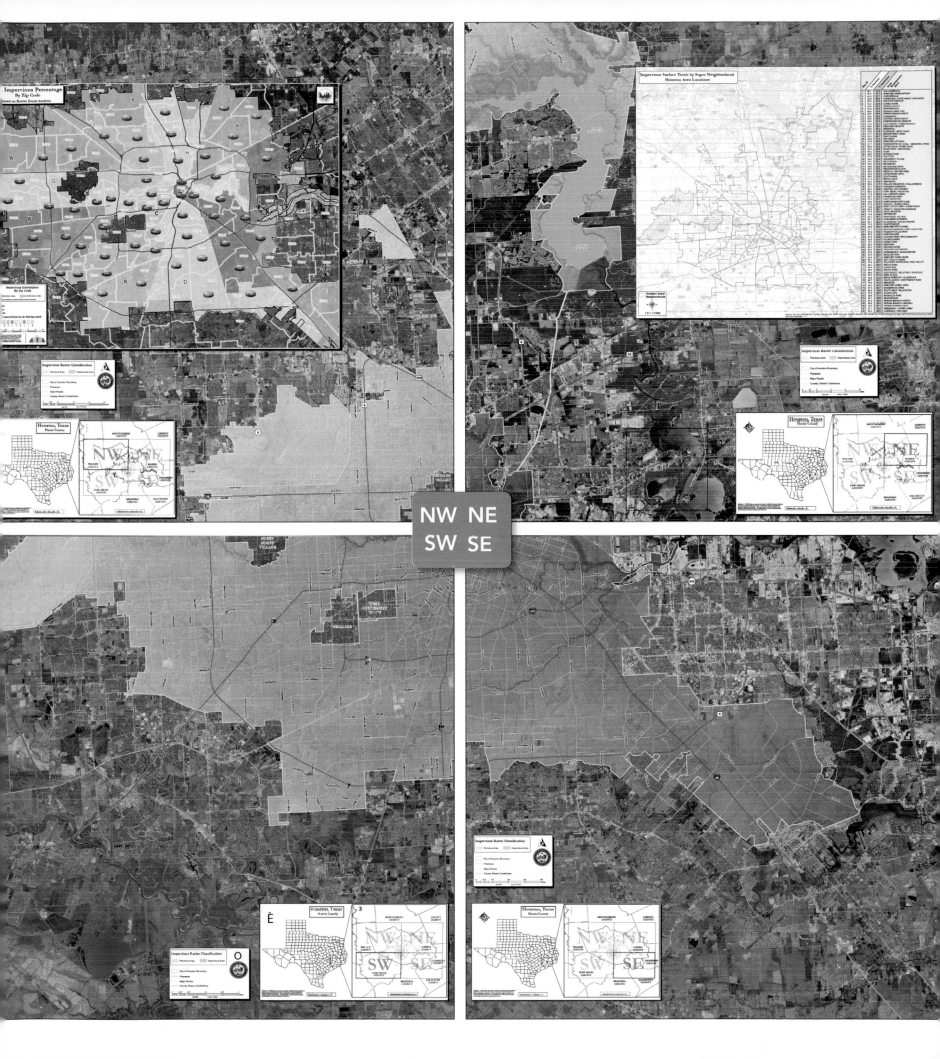

NW NE
SW SE

Live and Work Patterns: A GIS Study of the Maricopa County Trip Reduction Survey

Maricopa Association of Governments
Phoenix, Arizona, USA
By Peter Burnett, Shannon Acevedo, and Anubhav Bagley

Contact
Peter Burnett
pburnett@azmag.gov

Software
ArcGIS 10 for Desktop

Data Sources
Maricopa Association of Governments, Maricopa County Trip Reduction Program, MAG Employer Database

The Maricopa Association of Governments (MAG) is the designated metropolitan planning organization for transportation planning in the Maricopa County, Arizona, region. MAG has also been designated by the governor to serve as the principal planning agency for the region in a number of areas, one of which is air quality.

As part of this air quality planning, MAG works with the Maricopa County Trip Reduction Program (TRP). This program works with employers and schools and asks them to reduce single occupant vehicle (SOV) trips and/or miles traveled to the work site by 10 percent for a total of five years, and then 5 percent for three additional years, or until a 60 percent rate of SOV travel is reached. Progress is tracked through an annual commute survey of employer/school sites with fifty or more employees/students at each location. The results of the survey are used to develop an annual plan that commits the employer/school to implementing and documenting various strategies to reduce SOV trips or miles. In 2011, over 500,000 employees were surveyed at 3,000 employer sites.

The maps that were created show the types of analysis that MAG is able to conduct with data from the survey. Using the data, MAG is able to analyze commute patterns, the modes of transportation used, job industries, and employment center dynamics. All this helps MAG to develop and understand relationships between industries and occupations, which provides a better understanding of the labor market and travel patterns for different job centers within Maricopa County. This is essential in developing good regional plans for transportation and air quality.

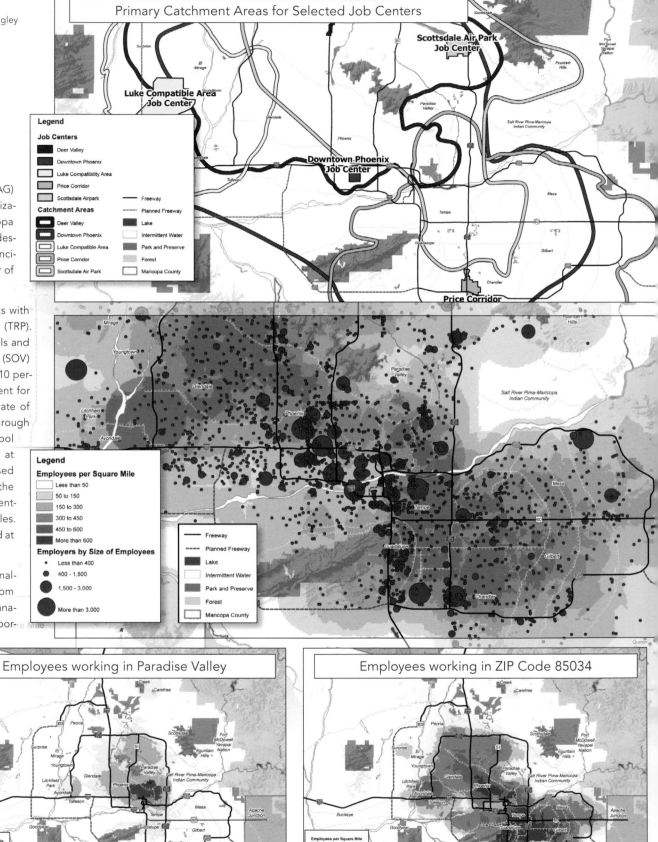

Primary Catchment Areas for Selected Job Centers

Employees working in Paradise Valley

Employees working in ZIP Code 85034

I MAP with a Purpose: Who We Are

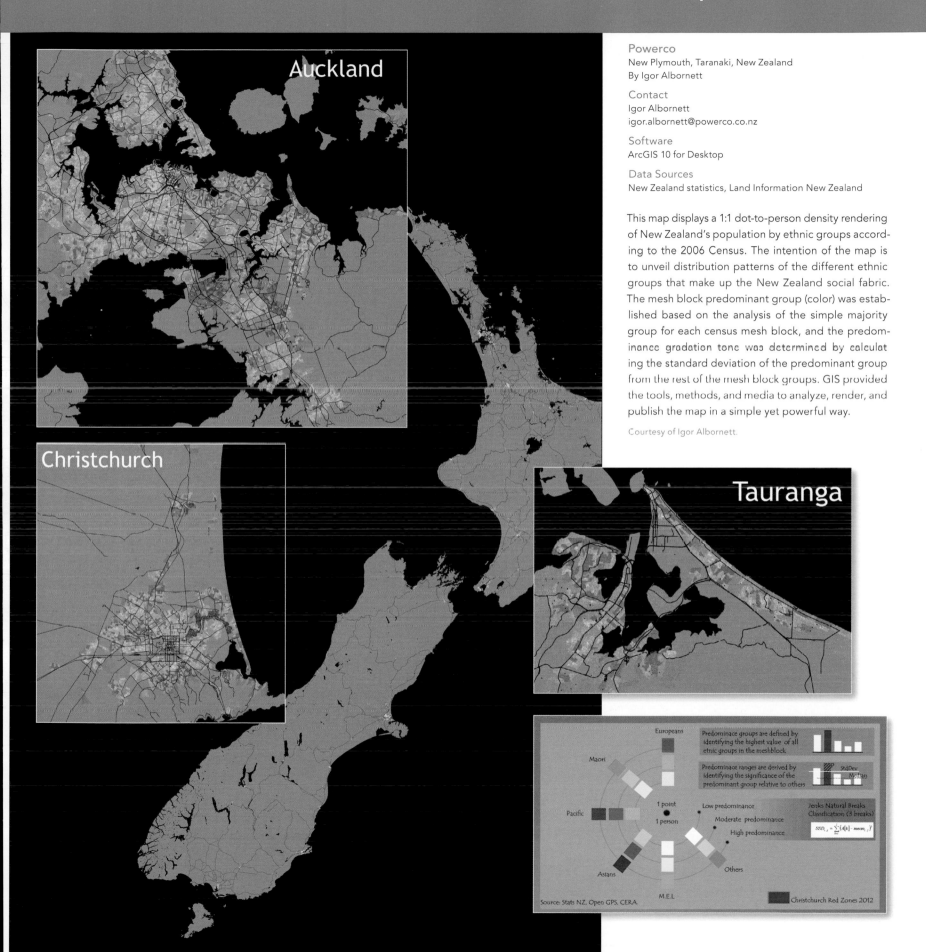

Auckland

Christchurch

Tauranga

Powerco
New Plymouth, Taranaki, New Zealand
By Igor Albornett

Contact
Igor Albornett
igor.albornett@powerco.co.nz

Software
ArcGIS 10 for Desktop

Data Sources
New Zealand statistics, Land Information New Zealand

This map displays a 1:1 dot-to-person density rendering of New Zealand's population by ethnic groups according to the 2006 Census. The intention of the map is to unveil distribution patterns of the different ethnic groups that make up the New Zealand social fabric. The mesh block predominant group (color) was established based on the analysis of the simple majority group for each census mesh block, and the predominance gradation tone was determined by calculating the standard deviation of the predominant group from the rest of the mesh block groups. GIS provided the tools, methods, and media to analyze, render, and publish the map in a simple yet powerful way.

Courtesy of Igor Albornett.

Lidar Applications for Property Assessment

Geographic Mapping Technologies Corp.
San Juan, San Juan, Puerto Rico
By Cesar Piovanetti

Contact
Cesar Piovanetti
cpiovanetti@gmtgis.com

Software
ArcGIS 10 for Desktop

Data Sources
National Oceanic and Atmospheric Administration, Puerto Rico Electric Power Authority, Junta de Planificación de Puerto Rico

One of the biggest advantages of lidar data is its versatility. It can be used for multiple government projects, appraisals, conservation of natural resources, and flood mapping, among many other uses. The purpose of this project was to demonstrate that lidar technology can be used to support the property assessment process. In this project, lidar data was used to accurately capture building footprints and the number of floors any given structure may have.

Courtesy of Geographic Mapping Technologies Corp.

Atlas of Access to Dialysis Centers in France

Agence de la Biomédecine (Biomedicine
Agency)
Saint-Denis La Plaine, Seine Saint-Denis, France
By Florian Bayer, Cécile Couchoud, Christian Jacquelinet,
and Mathilde Lassalle

Contact
Florian Bayer
florian.bayer@biomedecine.fr

Software
ArcGIS 10 for Desktop, ArcGIS Network Analyst, ArcGIS
Spatial Analyst, Adobe Illustrator, Scribus

Data Sources
REIN Agence de la Biomédecine 2011, CIAT-CSI
(SRTM http://srtm.csi.cgiar.org) 2010

The French Renal Epidemiology and Information Network (REIN) registry was created to contribute to the development and evaluation of health strategies, aiming at improving prevention and management of end-stage renal disease. In 2009, 37,500 patients were on dialysis, and 33,000 were living with a renal functioning graft. The national allocated budget for dialysis was about €4 billion during this time, roughly 3 percent of the French public health insurance system budget. The travel time between home and the dialysis center is a medico-economic and organizational indicator, the journey cost representing 13 percent of the hemodialysis budget. In 2009, 7.5 percent of patients took over 45 minutes to reach their dialysis center in France. In an economic environment of budget cuts, geographic approach of dialysis time access is a sensitive issue. In this context, the REIN registry

needed to find a simple, reliable, and generally applicable indicator of health care supply adequacy as a tool used to provide decision-making support for health planning.

All the 2009 patients have been geolocated with ArcGIS 10 Desktop. They were allocated to the nearest dialysis center, and the theoretical access time was calculated with ArcGIS Network Analyst and ArcGIS Spatial Analyst. Combined with demographic data, this approach can provide new documents for decision makers. It's also a powerful tool for territory diagnosis and for simulation to provide better health planning decisions such as the best location for a new dialysis center or for relocation and the effects on health care if two centers are grouped together.

Courtesy of Agence de la Biomédecine.

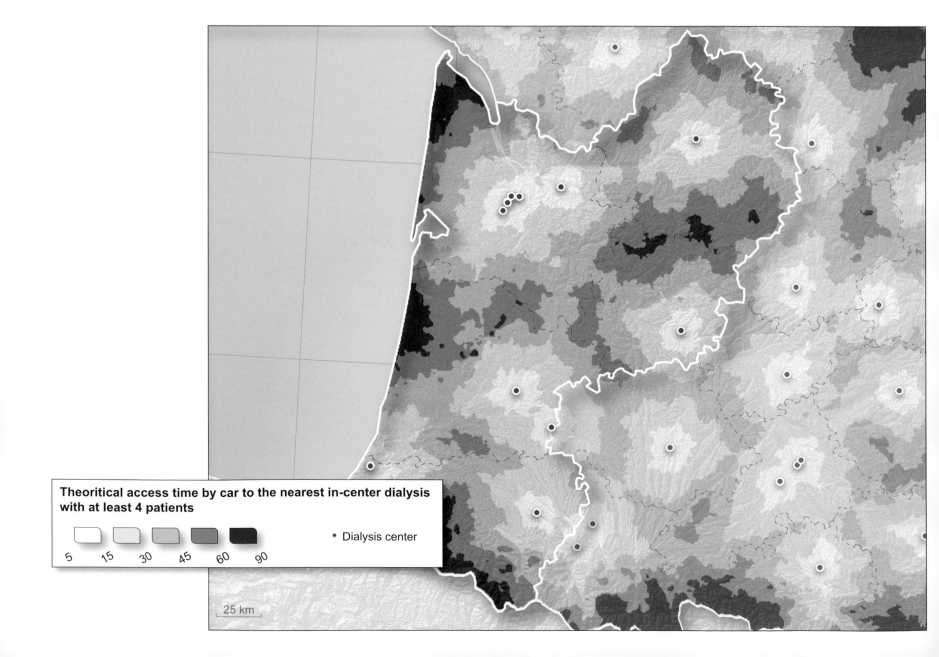

Theoritical access time by car to the nearest in-center dialysis with at least 4 patients

5   15   30   45   60   90

• Dialysis center

25 km

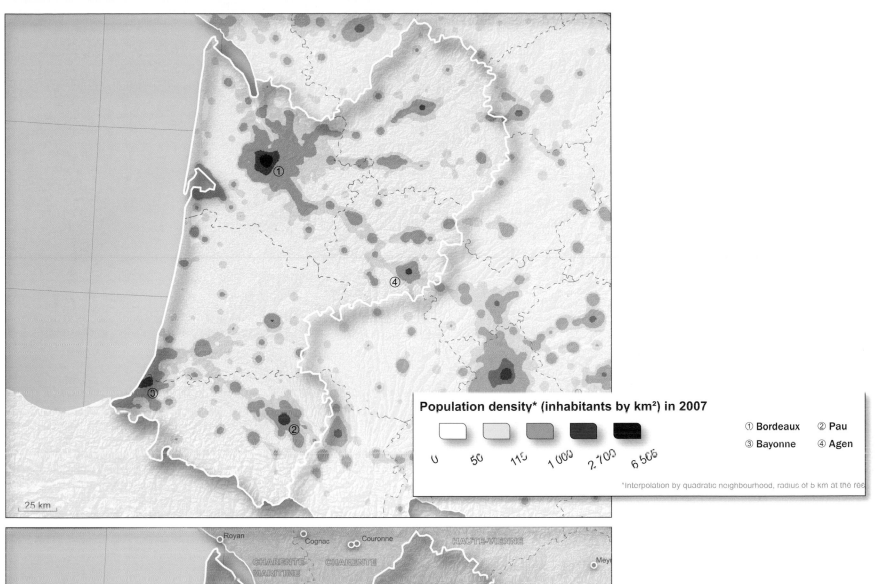

Population density* (inhabitants by km²) in 2007

0 50 110 1 000 2 700 6 505

① Bordeaux ② Pau
③ Bayonne ④ Agen

*Interpolation by quadratic neighbourhood, radius of 5 km at the rée...

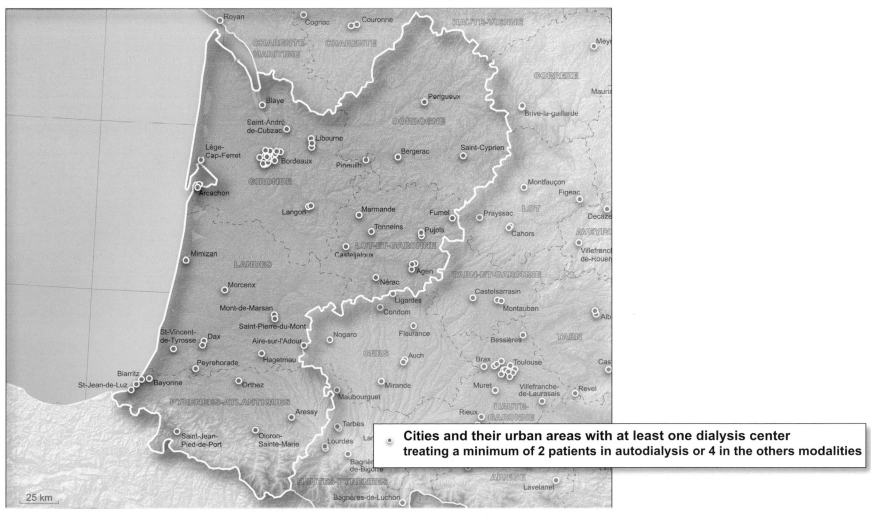

Royan · Cognac · Couronne · HAUTE-VIENNE
CHARENTE MARITIME · CHARENTE · Meyr
CORREZE
Blaye · Périgueux · Brive-la-gaillarde · Mauria
Lège-Cap-Ferret · Saint-André de-Cubzac · Libourne · DORDOGNE
Bordeaux · Pineuilh · Bergerac · Saint-Cyprien
GIRONDE · Montfauçon · Figeac
Arcachon · Langon · Marmande · Fumel · Prayssac · LOT · Decaze
Tonneins · Pujols · Cahors · AVEYRO
Mimizan · LANDES · Casteljaloux · LOT-ET-GARONNE · Villefranch de-Rouer
Morcenx · Nérac · Agen · TARN-ET-GARONNE
Mont-de-Marsan · Ligardes · Castelsarrasin
Saint-Pierre-du-Mont · Condom · Montauban · TARN
St-Vincent-de-Tyrosse · Dax · Aire-sur-l'Adour · Nogaro · Fleurance · Bessières · Alb
Peyrehorade · Hagetmau · GERS · Auch · Brax · Toulouse · Cas
Biarritz · Orthez · Mirande · Muret · Villefranche-de-Laurasais · Revel
St-Jean-de-Luz · Bayonne · PYRENEES-ATLANTIQUES · Maubourguet · Rieux · HAUTE-GARONNE
Aressy · Tarbes · Lan
Saint-Jean-Pied-de-Port · Oloron-Sainte-Marie · Lourdes
Bagnè de-Bigorre · ARIEGE · Lavelanet
HAUTES-PYRENEES · Bagnères-de-Luchon

Cities and their urban areas with at least one dialysis center
treating a minimum of 2 patients in autodialysis or 4 in the others modalities

25 km

25 km

Czech Lands within the Habsburg Monarchy during Seventeenth and Eighteenth Centuries

Czech Technical University in Prague
Prague, Czech Republic
By Pavel Seemann, Jiri Cajthaml, Tomas Janata, and
Ruzena Zimova

Contact
Jiri Cajthaml
jiri.cajthaml@fsv.cvut.cz

Software
ArcGIS 10 for Desktop

Data Source
Institute of History, Academy of Sciences of the Czech
Republic

This map was created as part of a 2013 atlas of Czech history. The atlas consists of over 300 historical maps published in cooperation with the Institute of History, Academy of Sciences of the Czech Republic (historical content), and Czech Technical University in Prague, Department of Mapping and Cartography (cartography).

The map shows the region of Central Europe during the seventeenth and eighteenth centuries. The main theme of the map is the Habsburg monarchy and its gained or lost provinces. The areas of the Czech lands are highlighted in purple, and the territory of the Habsburg monarchy before 1635 is depicted in orange. Hatched areas were gained by Habsburgs from 1699 to 1797, while green areas were occupied by the Ottoman Empire during the seventeenth century. The red line shows the Habsburg monarchy boundary in 1797 after the Campo Formio peace treaty.

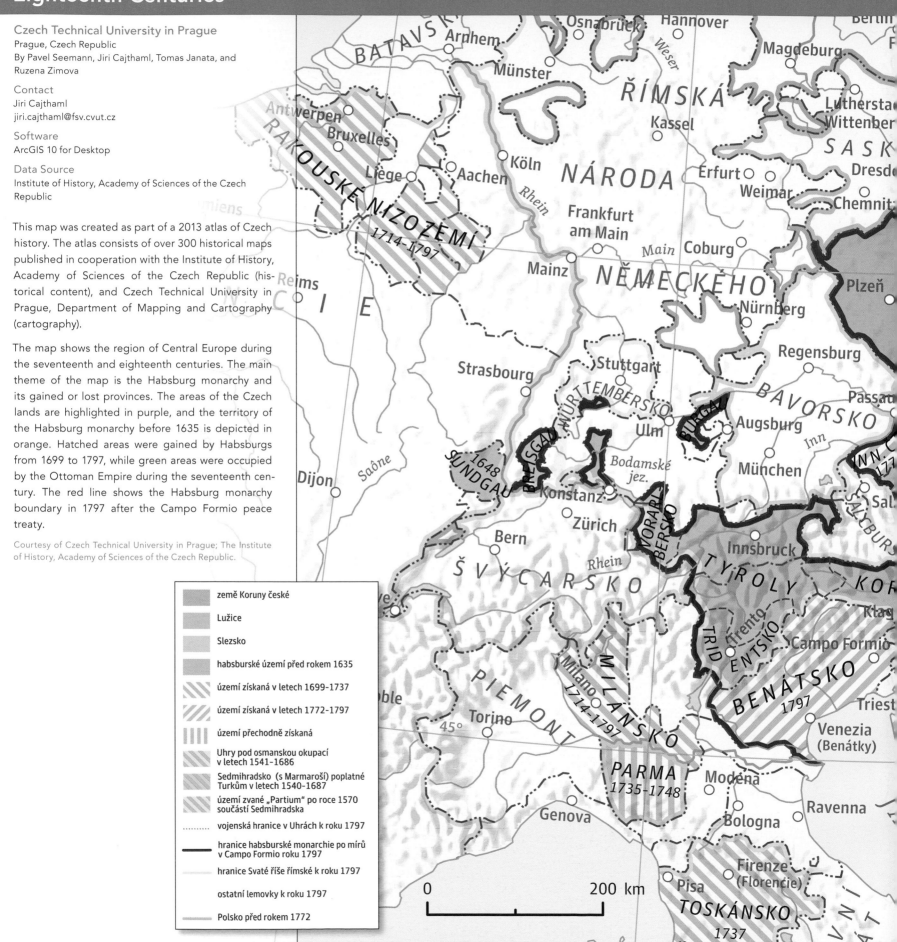

Legend:

- země Koruny české
- Lužice
- Slezsko
- habsburské území před rokem 1635
- území získaná v letech 1699–1737
- území získaná v letech 1772–1797
- území přechodně získaná
- Uhry pod osmanskou okupací v letech 1541–1686
- Sedmihradsko (s Marmaroší) poplatné Turkům v letech 1540–1687
- území zvané „Partium" po roce 1570 součástí Sedmihradska
- vojenská hranice v Uhrách k roku 1797
- hranice habsburské monarchie po míru v Campo Formio roku 1797
- hranice Svaté říše římské k roku 1797
- ostatní lemovky k roku 1797
- Polsko před rokem 1772

0   200 km

Geodesign for Restoration Planning in Disaster-Affected Regions

Ritsumeikan University
Kyoto, Kyoto, Japan
By Keiji Yano, Keigo Matsuoka, Yuzuru Isoda, Takashi Kirimura, Toshikazu Seto, and Tomoki Nakaya

Contact
Keiji Yano
yano@lt.ritsumei.ac.jp

Software
ArcGIS 10 for Desktop

Data Sources
Basic survey on natural environment (Ministry of the Environment), GISMAP for road (Hokkaido-Chizu Co., Ltd.), Esri Japan Corporation (Ishinomaki City, etc.), Estimated Tsunami Flooding Area (Pasco Co), Zmap TownII (Zenrin Co., Ltd.)

Restoration planning in regions damaged by the Great East Japan Earthquake on March 11, 2011, is an urgent issue. One year after the earthquake and tsunami, the infrastructure and economy of the disaster-affected regions are steadily recovering. Most of the disaster-affected local governments have formed reconstruction plans. However, they still need time to implement the plans. Geodesign is employed to change existing situations into preferred ones in terms of geographic space. As Carl Steinitz, the Harvard University professor, provides a framework of geodesign for landscape planning using GIS, his framework is being applied to support restoration planning in the disaster-affected regions.

For the City of Ishinomaki, Miyagi Prefecture, for example, a variety of its geospatial data is applied to the geodesign framework. After creating maps depicting attractive temporary housing and tsunami risks, the locations of actual temporary houses are evaluated to determine any conflicts between residential attractiveness and tsunami risk. Some temporary houses are located within high tsunami risk areas, while others are located within areas low in attractiveness for residences.

From a long-term standpoint, local governments can use the geodesign framework for constructing new permanent residences, new shopping centers, industrial locations, and so on, in disaster-affected areas.

Courtesy of Ritsumeikan University.

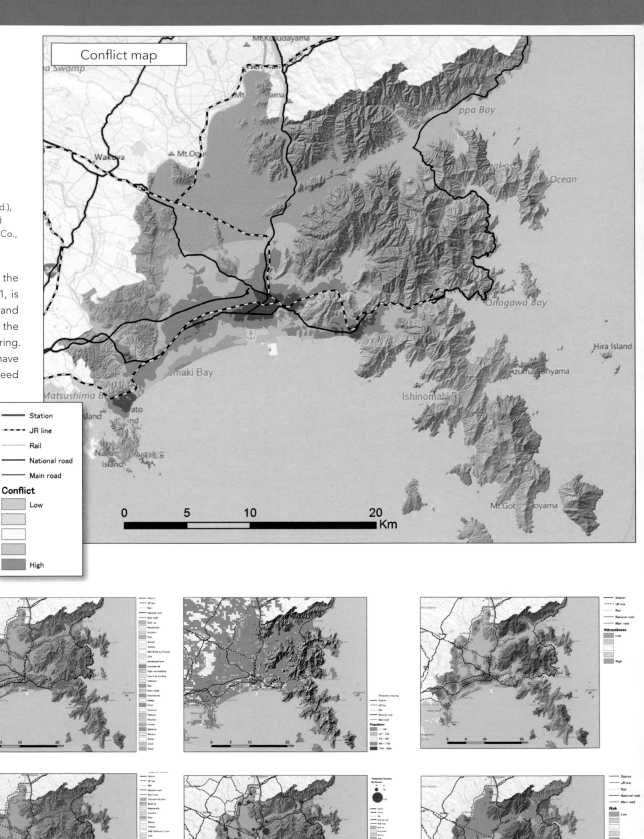

Measuring the Deterrent Effect of Poppy Eradication in the Helmand Food Zone

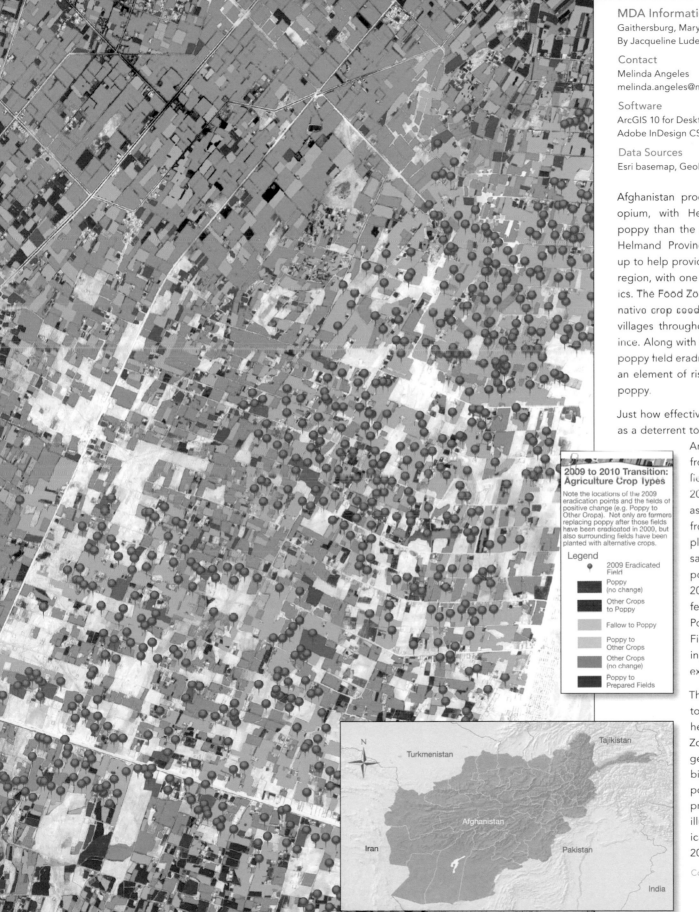

2009 to 2010 Transition: Agriculture Crop Types

Note the locations of the 2009 eradication points and the fields of positive change (e.g. Poppy to Other Crops). Not only are farmers replacing poppy after those fields have been eradicated in 2009, but also surrounding fields have been planted with alternative crops.

Legend

- 2009 Eradicated Field
- Poppy (no change)
- Other Crops to Poppy
- Fallow to Poppy
- Poppy to Other Crops
- Other Crops (no change)
- Poppy to Prepared Fields

MDA Information Systems, LLC
Gaithersburg, Maryland, USA
By Jacqueline Luders, Dan Meeks, and Melinda Angeles

Contact
Melinda Angeles
melinda.angeles@mdaus.com

Software
ArcGIS 10 for Desktop, Adobe Photoshop, Adobe Illustrator, Adobe InDesign CS3

Data Sources
Esri basemap, GeoEye-1, Alcis, USG

Afghanistan produces the majority of the world's opium, with Helmand Province cultivating more poppy than the rest of the country combined. The Helmand Provincial Reconstruction Team was set up to help provide stability and development in this region, with one of its key focuses on counternarcotics. The Food Zone Program distributes bags of alternative crop seeds and fertilizer among participating villages throughout the central region of the province. Along with incentives to plant alternative crops, poppy field eradication is part of this program to add an element of risk to farmers' decisions to cultivate poppy.

Just how effective is the eradication of poppy fields as a deterrent to growing poppy the following year? And how far-reaching is that impact from the location of the eradicated fields? Cultivation data from both 2009 and 2010 was compared to assess any patterns that might emerge from areas where eradication took place. Commercial high-resolution satellite data imaged during the peak poppy growing season in 2009 and 2010 was used in a semiautomated feature extraction process in which Poppy, Wheat, Other Crops, Fallow Fields, and Prepared Fields (signifying fields that were not poppy) were extracted.

The resulting census data was used to identify individual crop cultivation hectares within the Helmand Food Zone, and poppy density maps were generated. This data was then combined in ArcGIS with the locations of poppy fields eradicated in 2009 to produce a comprehensive analysis illustrating the effectiveness of eradication in Helmand Food Zone from 2009 to 2010 on a localized population.

Courtesy of MDA Information Systems, LLC

Agricultural Development 1850–2000

Nordpil
Stockholm, Sweden
By Hugo Ahlenius

Contact
Hugo Ahlenius
hugo.ahlenius@nordpil.com

Software
ArcGIS for Desktop, Adobe Illustrator, Global Mapper

Data Sources
Klein Goldewijk, Kees, Arthur Beusen, Martine de Vos, and Gerard van Drecht. 2011. "The HYDE 3.1 Spatially Explicit Database of Human Induced Land Use Change over the Past 12,000 Years." Global Ecology and Biogeography 20 (1): 73–86. DOI: 10.1111/j.1466-8238.2010.00587.x.

This map was prepared for a scientific paper on water management and agriculture. The development and expansion of agriculture have drastically changed the face of this planet. In the Anthropocene (the epoch of humans capable of changing earth's atmosphere), most of the productive regions in the world were used for grazing land and cropland. The paper concluded that humans may have shifted from living in the Holocene (emergence of agriculture) into the Anthropocene before the Industrial Revolution in light of the sheer size and magnitude of some historial land-use changes, including the fourtheenth century Black Death that depopulated Europe and the aftermath of the colonization of the Americas in the sixteenth century.

Courtesy of Hugo Ahlenius and Mats Lannerstad, Stockholm Environment Institute.

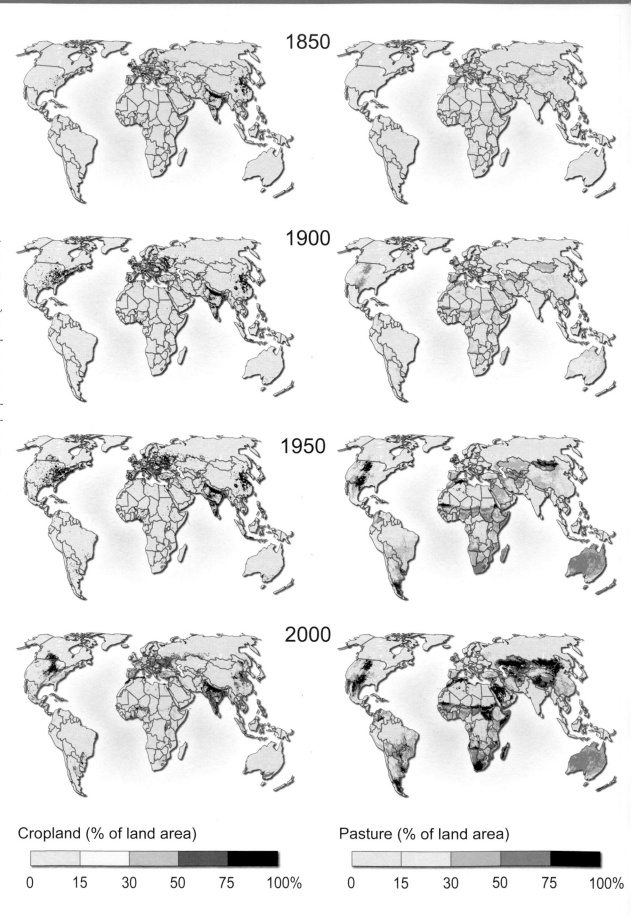

1850

1900

1950

2000

Cropland (% of land area)

0   15   30   50   75   100%

Pasture (% of land area)

0   15   30   50   75   100%

Yuma Proving Ground Potential Fuel Load and Fire Modeling

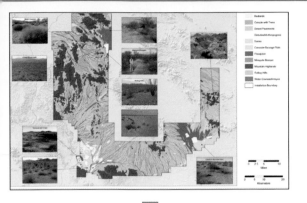

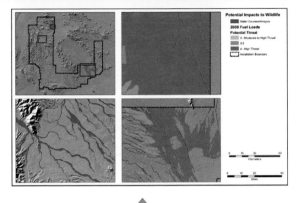

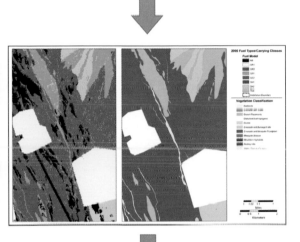

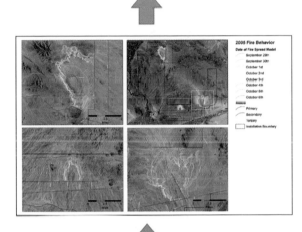

KAYA Associates, Inc.
Huntsville, Alabama, USA
By Joshua Chris, David Wright, and Wesley Norris

Contact
Wesley Norris
norrisw@kayacorp.com

Software
ArcGIS 10 for Desktop, FlamMap, FireFamilyPlus, FARSITE, LANDFIRE, FAMWEB

Data Sources
Yuma Proving Ground, Desert Research Institute, Esri, US Department of Agriculture Forest Service Rocky Mountain Field Station, KAYA Associates, Inc.

Yuma Proving Ground initially tasked KAYA with developing an installation-wide vegetation classification. This required extensive fieldwork in a nearly yearlong project working with professionals from the University of Arizona. The flowchart outlines the processes, software, and data sources used to support the fire fuel potential maps and fire behavior modeling.

The vegetation layer served as a baseline for all fire fuel and fire modeling efforts and was correlated to a fire fuel class using the Scott and Burgan 40 Model. KAYA worked closely with the Rocky Mountain Field Station (RMFS) and implemented its custom software packages including FARSITE and FlamMap. Initial output was a fuel load potential threat layer that identifies areas that should be monitored during the dry season. Threats were modeled for average precipitation years (1998) and above-average precipitation years (2005). The fuel threat layer was then used to develop a fire model for both the 1998 and 2005 conditions. Ignition points were chosen based on operational layers such as firing lines, gun positions, and impact areas. Additional ignition points have since been used for training exercises for first responders.

The combination of these efforts can be used for a variety of analyses. The Wildlife Impact map shows areas with a high threat for fire, overlaid with vegetated wash areas that are typically the most densely populated with local fauna. This can be used by the environmental staff to mitigate the damage that may be caused to local habitats in the event of a wildfire.

Courtesy of Yuma Proving Ground, KAYA Associates, Inc.

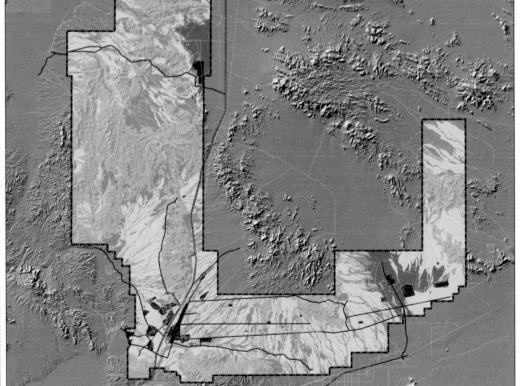

2005 Fuel Loads
Potential Threat

- 0 - Non-Burnable
- 0.5
- 1 - Low Threat
- 1.5
- 2 - Moderate Threat
- 2.5
- 3 - Moderate to High Threat
- 3.5
- 4 - High Threat
- Primary
- Secondary
- Tertiary
- Installation Boundary

Prescribed Burn Season

US Army Fort Bragg
Fort Bragg, North Carolina, USA
By Susan Pulsipher

Contact
Susan Pulsipher
susan.e.pulsipher.ctr@mail.mil

Software
ArcGIS 10 for Desktop

Data Source
Fort Bragg Forestry Branch

The Forestry Branch at the Fort Bragg US Army installation, North Carolina, conducts prescribed burns on approximately 53,000 acres of woodland and training lands every year. Prescribed burns are used to maintain drop zones and ranges for troop training, to maintain the longleaf pine ecosystem, and to reduce the risks from wildfires.

Fire management personnel, with assistance from the installation GIS specialists, produce a series of base-wide maps for every stage of the annual burn program. Maps are used to develop and refine the upcoming season's burn plan; to identify which firebreak roads to inspect, repair, and grade; to prepare the ground for specific burns; and to illustrate progress during the season. Burns are coordinated with military trainers, personnel working with endangered species, timber operators, and other users of the training lands. These maps provide a frame of reference and highlight important features, such as the location of red-cockaded woodpecker (RCW) cavity trees.

This map shows the FY 2012 burn plan and status of burn blocks. It is a working document used by administrators, illustrating what burn activities have been accomplished. Fort Bragg is divided into burn block units, and each has an assigned burn frequency. All the blocks scheduled for burning in a year are color coded according to whether the burn will be in dormant season (fuel reduction) or in growing season (habitat improvement). The GIS data is updated daily and reflected in maps used for status reporting.

Courtesy of US Army Fort Bragg Environment Division and the Office of the Assistant Chief of Staff for Installation Management Installation Geospatial Information and Services.

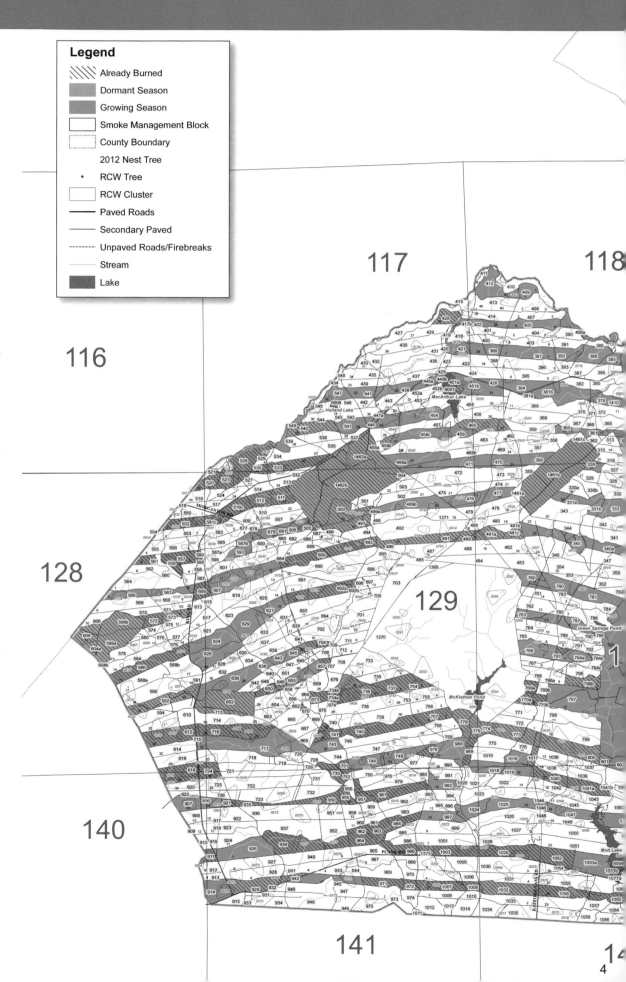

Legend

⫽⫽⫽	Already Burned
▨	Dormant Season
▩	Growing Season
☐	Smoke Management Block
⬚	County Boundary
☐	2012 Nest Tree
•	RCW Tree
☐	RCW Cluster
▬▬	Paved Roads
———	Secondary Paved
-------	Unpaved Roads/Firebreaks
——	Stream
▨	Lake

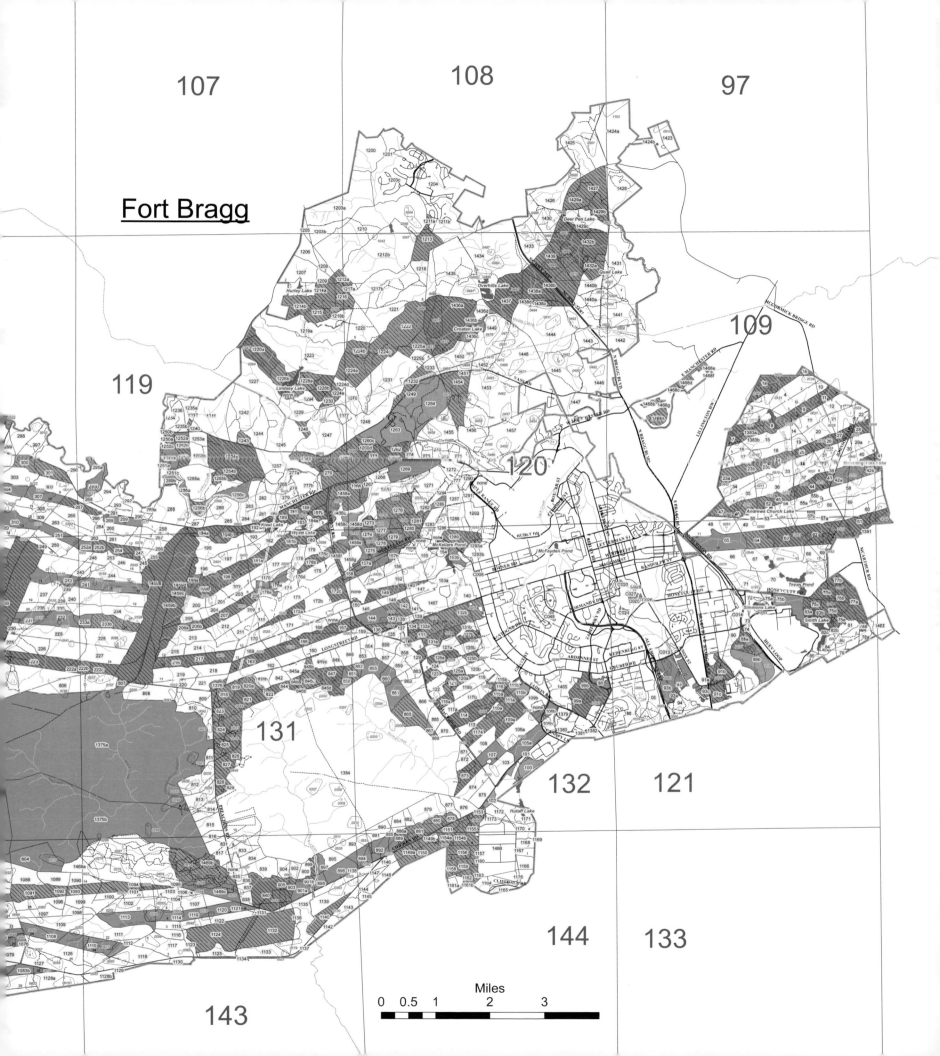

Fort Bragg

Pantropical Aboveground Woody Biomass

Woods Hole Research Center
Falmouth, Massachusetts, USA
By Greg Fiske

Contact
Greg Fiske
gfiske@whrc.org

Software
ArcGIS for Desktop

Data Source
Woods Hole Research Center

Scientists at the Woods Hole Research Center and their collaborators have generated the first high-resolution dataset showing the distribution and magnitude of aboveground woody biomass in the tropics. Using a combination of colocated field measurements, lidar observations, and imagery recorded from the Moderate Resolution Imaging Spectroradiometer, Woods Hole researchers produced national level maps showing the amount and spatial distribution of aboveground carbon.

Courtesy of Woods Hole Research Center.

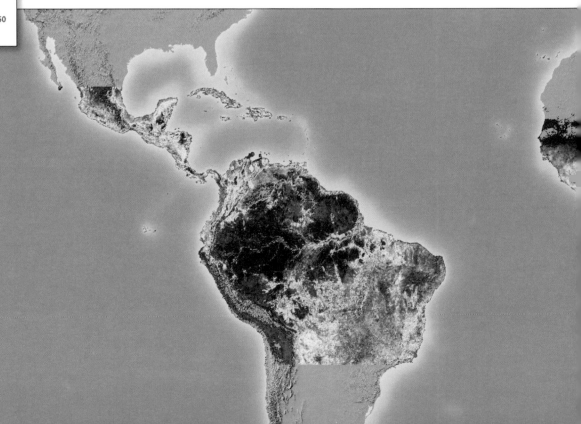

Predicted Aboveground Woody Biomass

| 0 | 10 | 100 | 175 | 250 | >450 |

tonnes per hectare

Geochemistry and Geophysics Field Maps Used during the USGS 2011 Field Season in Southwest Alaska

US Geological Survey
Denver, Colorado, USA
By Stuart A. Giles

Contact
Stuart Giles
sgiles@usgs.gov

Software
ArcGIS 10.1 for Desktop

Data Source
USGS, *National Atlas of the United States*

The US Geological Survey (USGS) has been studying a variety of geochemical and geophysical assessment techniques for concealed mineral deposits. The 2011 field season for this project took place in southwest Alaska, northeast of Bristol Bay between Dillingham and Iliamna Lake. Four maps were created for the geochemistry and geophysics teams to use during field activities.

GIS was used to provide each team with maps that displayed specific information necessary to conduct fieldwork. The geophysics team required regional maps, with layers that included existing aeromagnetic data, digital elevation surfaces, bedrock geology, and critical boundaries such as native lands and national parks. The geochemistry team required maps at two scales: a moderate-scale map showing area bedrock geology, digital elevation, and major hydrology and a large-scale map showing current mine claim boundaries, detailed local hydrology, and elevation colored and shaded to emphasize small differences in terrain.

Survey lines and sampling sites were chosen based on factors specific to each team. The geophysics team required dry-land sites in a series of northwest-trending lines across the region with sensitivity to private lands. The geochemistry team required accurate locations of ponds and streams for water sampling and glacial moraines for till sampling. Coordinates of selected sites were calculated and placed in a project GIS, plotted on the field maps, then uploaded as waypoints into team GPS units for use in the field.

Courtesy of US Geological Survey, Alaska Department of Natural Resources.

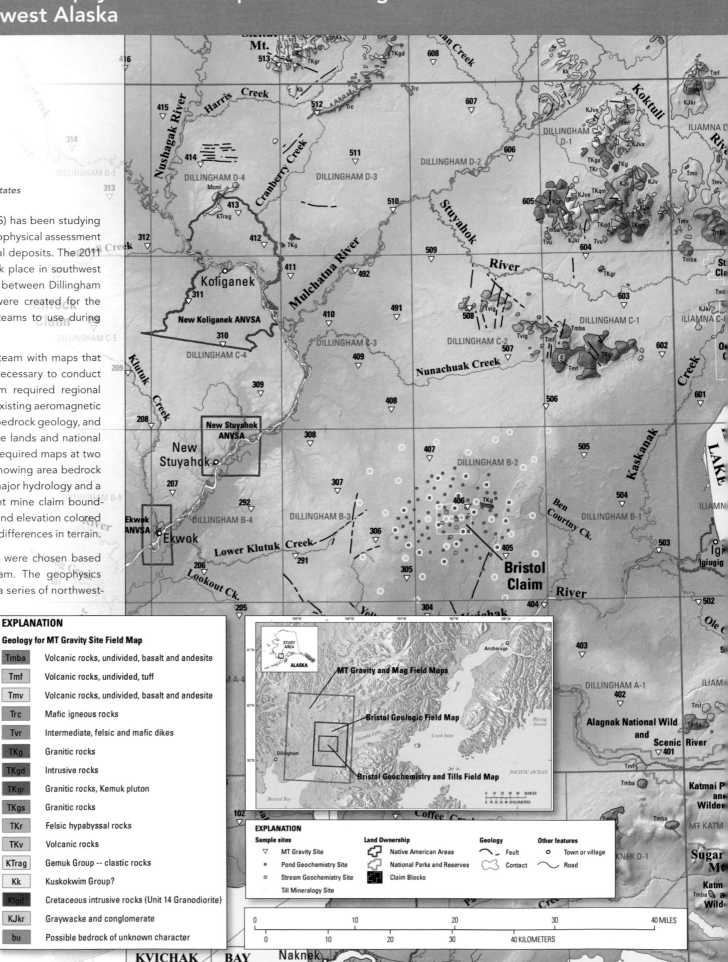

EXPLANATION

Geology for MT Gravity Site Field Map

Tmba	Volcanic rocks, undivided, basalt and andesite
Tmf	Volcanic rocks, undivided, tuff
Tmv	Volcanic rocks, undivided, basalt and andesite
Trc	Mafic igneous rocks
Tvr	Intermediate, felsic and mafic dikes
TKg	Granitic rocks
TKgd	Intrusive rocks
TKgr	Granitic rocks, Kemuk pluton
TKgs	Granitic rocks
TKr	Felsic hypabyssal rocks
TKv	Volcanic rocks
KTrag	Gemuk Group -- clastic rocks
Kk	Kuskokwim Group?
KIgd	Cretaceous intrusive rocks (Unit 14 Granodiorite)
KJkr	Graywacke and conglomerate
bu	Possible bedrock of unknown character

EXPLANATION

Sample sites
▽ MT Gravity Site
• Pond Geochemistry Site
▢ Stream Geochemistry Site
Till Mineralogy Site

Land Ownership
Native American Areas
National Parks and Reserves
Claim Blocks

Geology
Fault
Contact

Other features
○ Town or village
Road

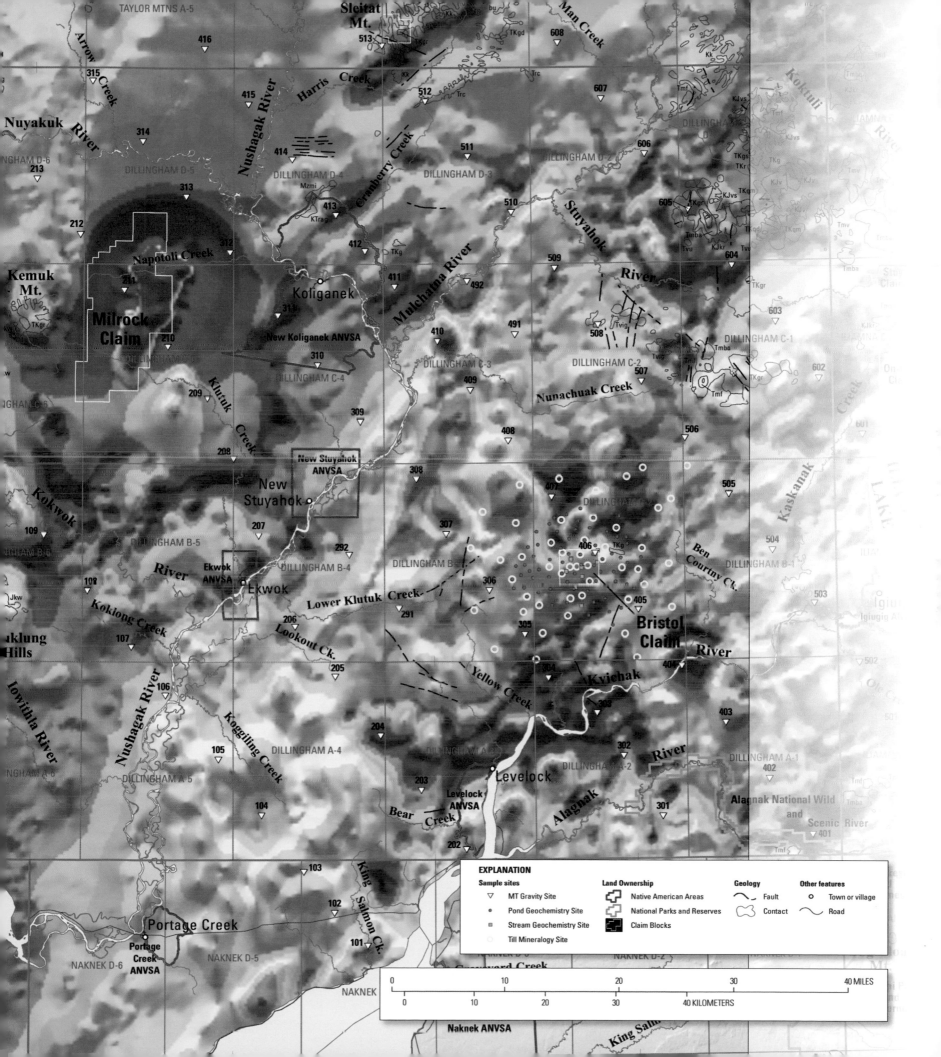

EXPLANATION

Sample sites
- ▽ MT Gravity Site
- • Pond Geochemistry Site
- ▪ Stream Geochemistry Site
- ◎ Till Mineralogy Site

Land Ownership
- Native American Areas
- National Parks and Reserves
- Claim Blocks

Geology
- Fault
- Contact

Other features
- ◦ Town or village
- Road

0   10   20   30   40 MILES

0   10   20   30   40 KILOMETERS

Geochemistry and Geophysics Field Maps Used during the USGS 2011 Field Season in Southwest Alaska

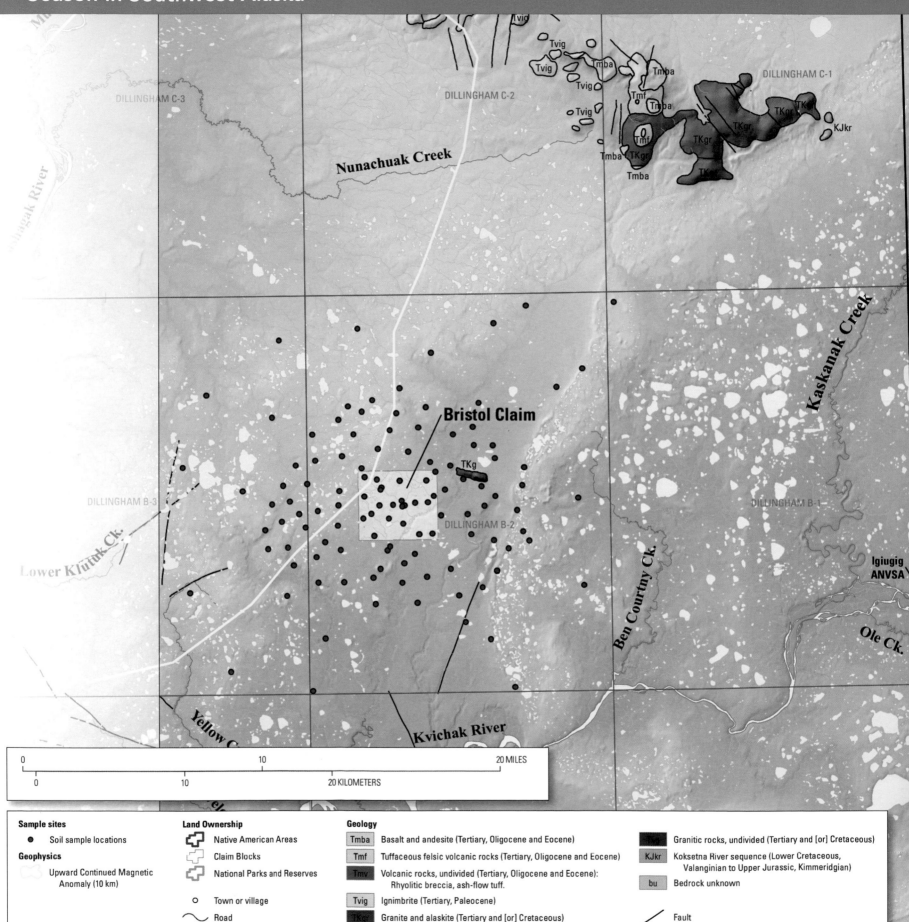

Sample sites
- Soil sample locations

Geophysics
Upward Continued Magnetic Anomaly (10 km)

Land Ownership
- Native American Areas
- Claim Blocks
- National Parks and Reserves
- ○ Town or village
- Road

Geology

Tmba — Basalt and andesite (Tertiary, Oligocene and Eocene)

Tmf — Tuffaceous felsic volcanic rocks (Tertiary, Oligocene and Eocene)

Tmv — Volcanic rocks, undivided (Tertiary, Oligocene and Eocene): Rhyolitic breccia, ash-flow tuff.

Tvig — Ignimbrite (Tertiary, Paleocene)

TKgr — Granite and alaskite (Tertiary and [or] Cretaceous)

TKg — Granitic rocks, undivided (Tertiary and [or] Cretaceous)

KJkr — Koksetna River sequence (Lower Cretaceous, Valanginian to Upper Jurassic, Kimmeridgian)

bu — Bedrock unknown

— Fault

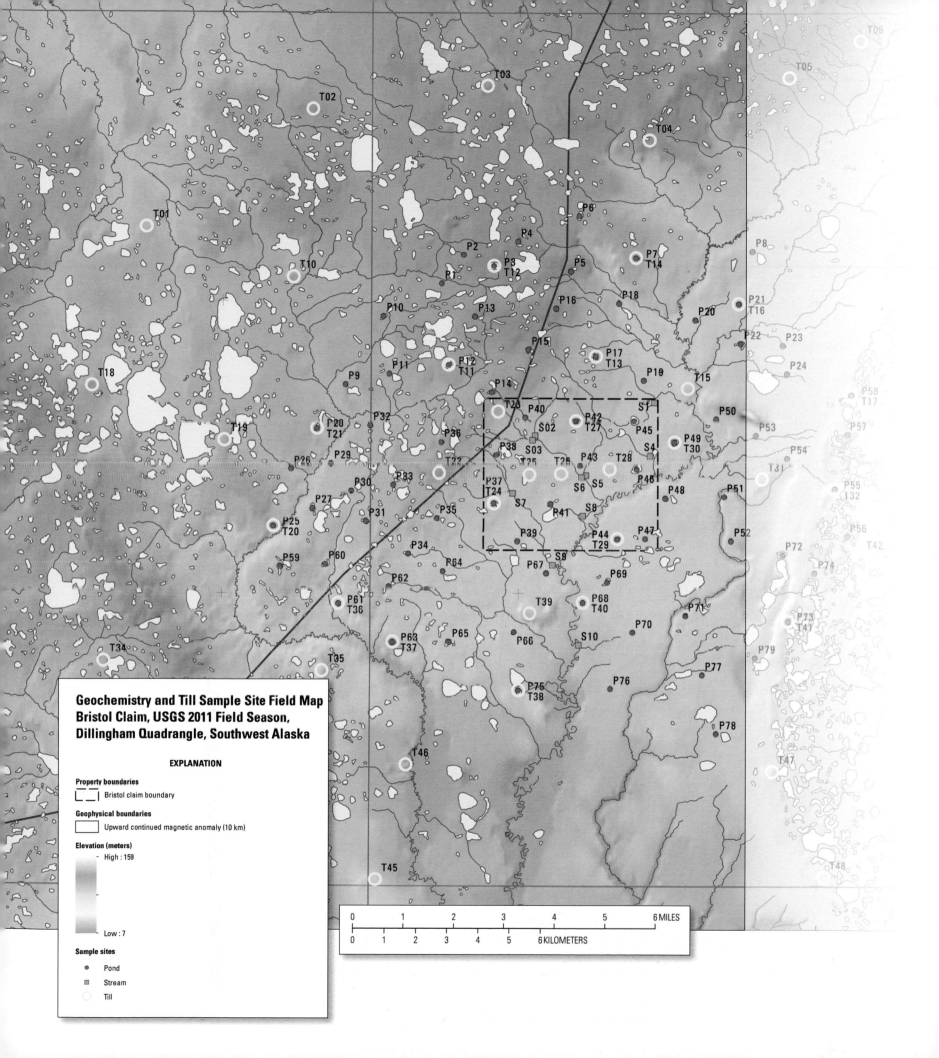

**Geochemistry and Till Sample Site Field Map
Bristol Claim, USGS 2011 Field Season,
Dillingham Quadrangle, Southwest Alaska**

EXPLANATION

Property boundaries

┌ ─ ┐
└ ─ ┘ Bristol claim boundary

Geophysical boundaries

□ Upward continued magnetic anomaly (10 km)

Elevation (meters)

High : 159

Low : 7

Sample sites

● Pond

■ Stream

◎ Till

| 0 | 1 | 2 | 3 | 4 | 5 | 6 MILES |

| 0 | 1 | 2 | 3 | 4 | 5 | 6 KILOMETERS |

Darfield Earthquake: A Study of Aftershock Epicenter Locations

San Diego Mesa College
San Diego, California, USA
By Karen Chadwick

Contact
Karen Chadwick
karenmchadwick@hotmail.com

Software
ArcGIS 10 Desktop, ArcGIS Geostatistical Analyst

Data Sources
GNS Science, Earthquake Commission, Statistics New Zealand, University of Otago, New Zealand Open GPS Maps, Peter Scott, Land Information New Zealand, Ollivier & Co, National Institute of Water and Atmospheric Research, Geographx, koordinates.com

The magnitude 7.1 Darfield earthquake near Christchurch, New Zealand, in September 2010 was surprising in its lack of casualties; there were no direct fatalities, despite a focal depth of only 11 kilometers. However, an aftershock of the Darfield earthquake, the magnitude 6.3 Christchurch earthquake in February 2011, resulted in 185 fatalities. Both earthquakes caused significant damage in Christchurch and nearby townships.

This study focused on over 7,500 earthquakes of magnitude 2.5 or greater occurring between September 4, 2010, and March 30, 2012, and was undertaken for course work in GIS at San Diego Mesa College. Generally, most earthquake aftershocks can be expected to occur along the main fault rupture or associated faults. The intent of the study was to determine whether the aftershocks of the Darfield earthquake, divided into four groups by time, occurred with the same spatial distribution. Spatial statistics indicated that the aftershock groups did not have the same spatial distribution: the centers were moving eastward, and the distributions were becoming elongated over time. Densities for each of the four aftershock groups also showed different distributions.

Courtesy of Karen Chadwick.

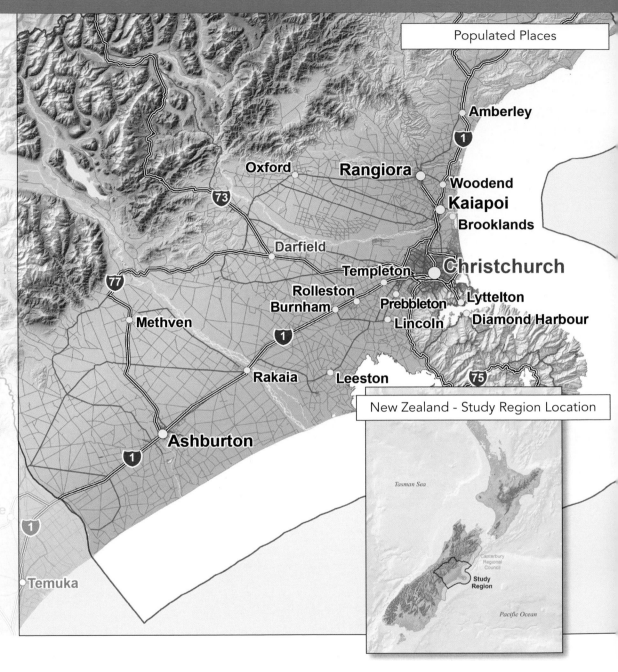

Populated Places

New Zealand - Study Region Location

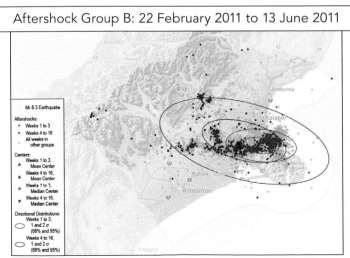

Aftershock Group B: 22 February 2011 to 13 June 2011

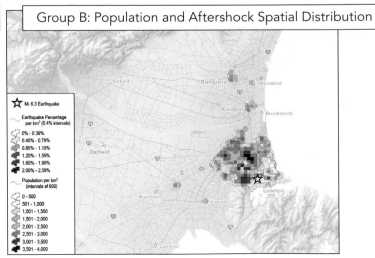

Group B: Population and Aftershock Spatial Distribution

National Oceanic and Atmospheric Administration's National Geodetic Survey

Silver Spring, Maryland, USA
By Brian Shaw and Dan Roman

Contact
Brian Shaw
brian.shaw@noaa.gov

Software
ArcGIS 10 for Desktop, Adobe Photoshop CS5

Data Sources
US Geological Survey, National Oceanic and Atmospheric Administration's GEOID12A, Esri, US Census

The National Oceanic and Atmospheric Administration's (NOAA) National Geodetic Survey's (NGS) mission is "to define, maintain, and provide access to the National Spatial Reference System (NSRS) to meet [the United States's] economic, social, and environmental needs." The NSRS is a consistent coordinate system that defines latitude, longitude, height, scale, gravity, and orientation throughout the United States and its territories. NGS dates back to the original Survey of the Coast established by President Thomas Jefferson in 1807. It is primarily tasked with determining where things are located and making sure that the American public can readily access that positioning for all commercial, scientific, and legal concerns.

GEOID12A is a principal component of the NSRS and is a datum conversion tool that maps the difference between the ellipsoid surface of the North American Datum of 1983 (NAD 83) and the North American Vertical Datum of 1988 (NAVD 88). Both NAD83 and NAVD88 are principal components of the NSRS and are the legal datums of the United States. Positions in the NAD83 are easily obtained using GPS and other Global Navigation Satellite System technology, but to obtain accurate elevations, a geoid model (GEOID12A) is needed to convert ellipsoid heights into orthometric heights (elevation).

GEOID12A is developed as a grid on a 1 arc-minute interval (approximately 2 kilometers). NGS continues to refine the geoid models as more data is accumulated, creating better models of the dynamic world. Since the earth's surface is infinitely complex, creating generalized models of it is needed to limit the detail. This map is a useful tool to help educate the public about GEOID12A and to provide insight on what a geoid is and how it should be used.

Courtesy of National Oceanic and Atmospheric Administration.

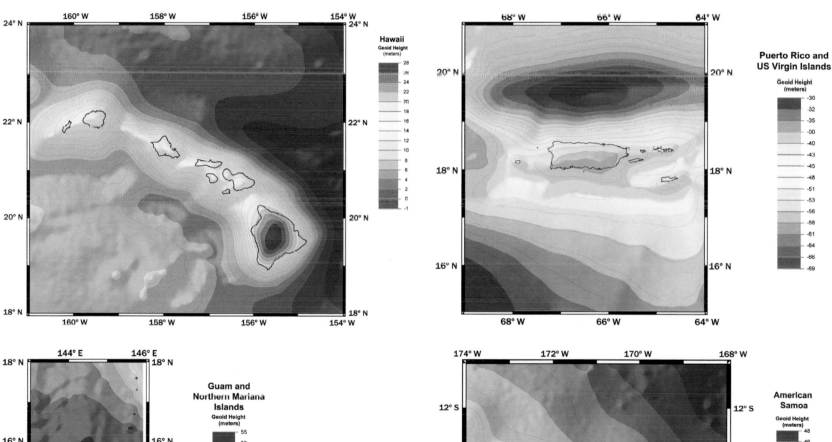

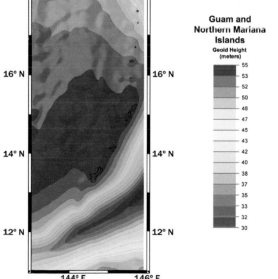

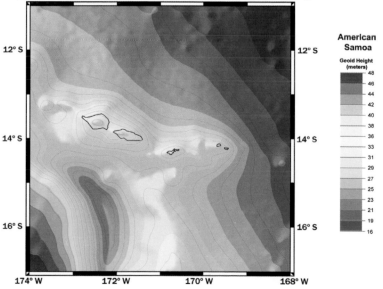

Geologic Map of Io

US Geological Survey
Flagstaff, Arizona, USA
By David Williams

Contact
Trent Hare
thare@usgs.gov

Software
ArcGIS Desktop 9.3, Adobe Illustrator

Data Sources
Voyager 1 (March 1979), Voyager 2 (July 1979), Hubble Space Telescope (1990–present), Galileo (1996–2001), Cassini (December 2000), and New Horizons (February 2007)

The National Aeronautics and Space Administration's (NASA) Voyager and Galileo datasets were merged to enable the characterization of the whole surface of the satellite at a consistent resolution. The US Geological Survey produced a set of four global mosaics of Io in visible wavelengths at a spatial resolution of 1 kilometer/pixel, released in February 2006, which were used as basemaps for this new global geologic map. Much has been learned about Io's volcanism, tectonics, degradation, and interior since the Voyager flybys, primarily during and following the Galileo Mission at Jupiter (December 1995–September 2003), and the results have been summarized in books published after the end of the Galileo Mission. This new understanding assists in map unit definition and in providing a global synthesis of Io's geology.

Courtesy of US Geological Survey, NASA.

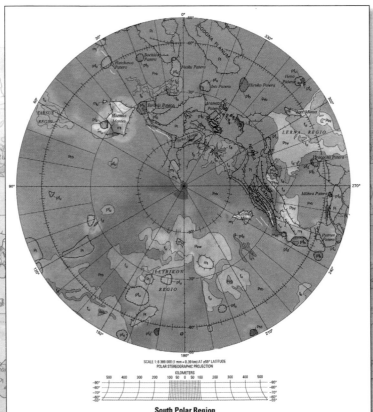

South Polar Region

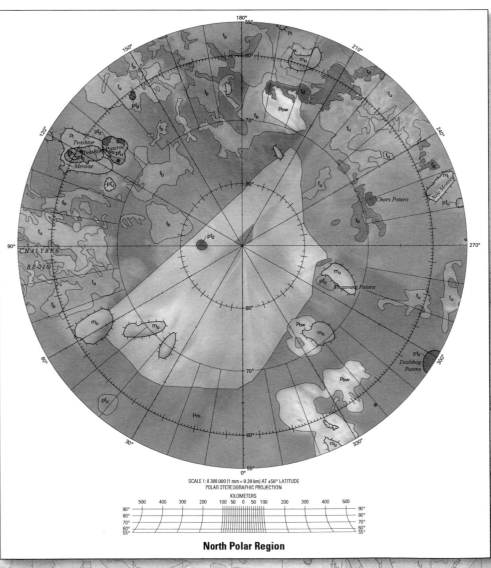

North Polar Region

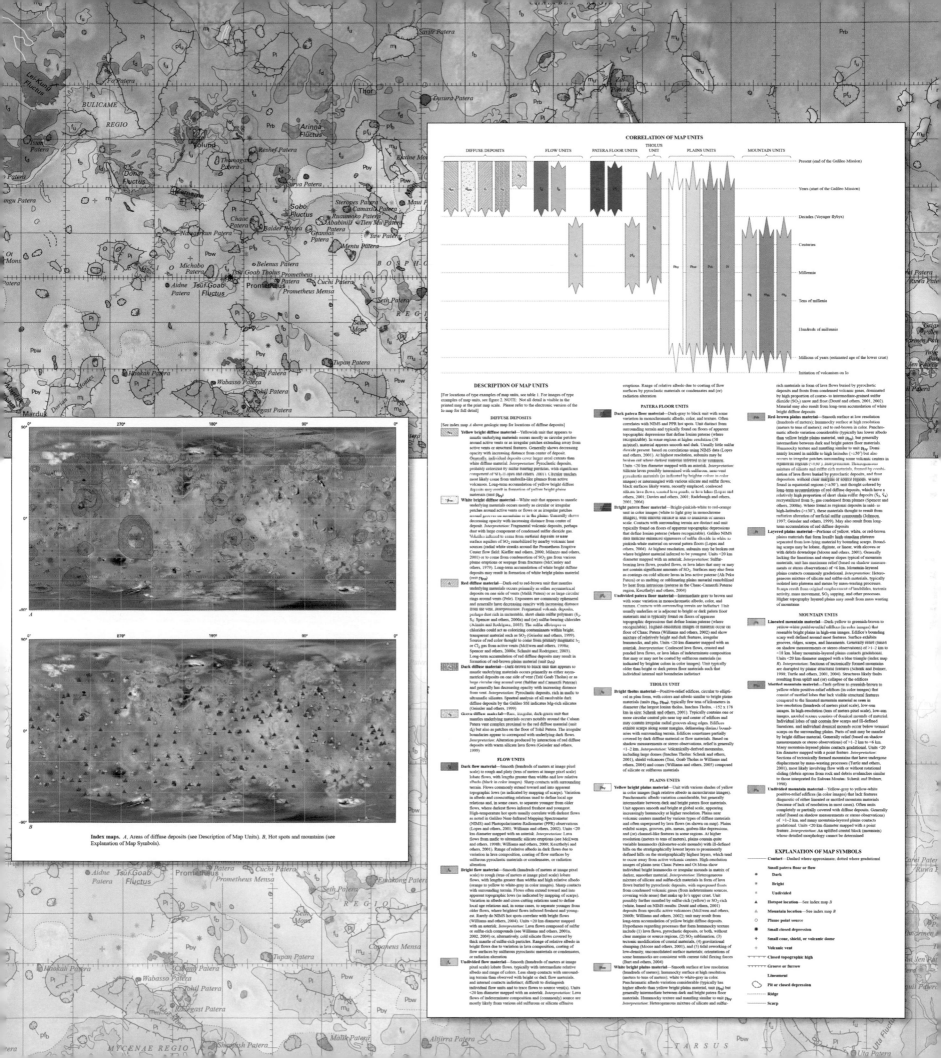

Index maps. *A*, Areas of diffuse deposits (see Description of Map Units). *B*, Hot spots and mountains (see Explanation of Map Symbols).

Virtual Astronaut Developed Based on GIS Tools and Unity: A Prototype Study at Santa Maria Crater on Mars

Washington University in St. Louis
St. Louis, Missouri, USA
By Jue Wang, Keith J. Bennett, and Raymond E. Arvidson

Contact
Jue Wang
wang@wunder.wustl.edu

Software
ArcGIS 10 for Desktop, Unity

Data Sources
Planetary data from NASA's orbital and landed missions

The Virtual Astronaut (VA) is an interactive Mars 3D environment created using multisource and multiinstrument data from orbital and landed missions. The VA provides virtual reality experience in a scene with imaging, compositional, and mineralogical details. It allows a virtual astronaut to explore the scene from any perspective, including a rover's. A prototype study of the VA was taken at Santa Maria Crater by the Geosciences Node of the National Aeronautics and Space Administration's (NASA) Planetary Data System (PDS), which archives and distributes digital data related to the study of the terrestrial planetary bodies such as Mars, Mercury, Venus, and Earth's moon. Santa Maria is an impact crater located at 2.172°S, 5.445°W in Meridiani Planum, visited by NASA Mars Exploration Rover *Opportunity*. The crater is approximately 90 meters in diameter and 16 meters in height.

The VA was created using ArcGIS and other tools to build a texture model, which includes several image mosaics overlaid on a 3D terrain model. The image mosaics have High Resolution Imaging Science Experiment (HiRISE) orbital data acquired from the Mars Reconnaissance Orbiter mission, as well as high-quality Panoramic Camera (Pancam) multispectral, Navigation Camera (Navcam), and Microscopic Imager (MI) ground-based data acquired by *Opportunity*. The VA was built on the Unity3D Game Engine with interactive functions based on customization of standard Unity modules. The figures that follow show multiple perspectives of the VA and screen shots of its interactive functions, including target observation, astronaut free walk, and dimension measurement of a feature.

Courtesy of Jue Wang, Keith J. Bennett, and Raymond E. Arvidson at NASA's PDS Geosciences Node.

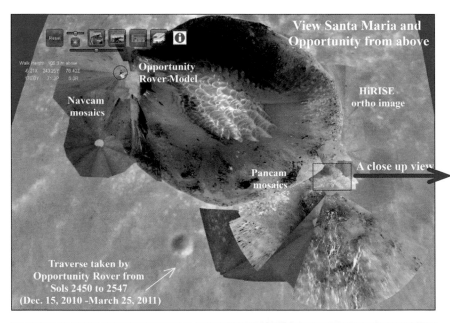

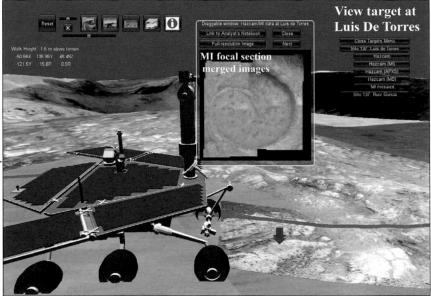

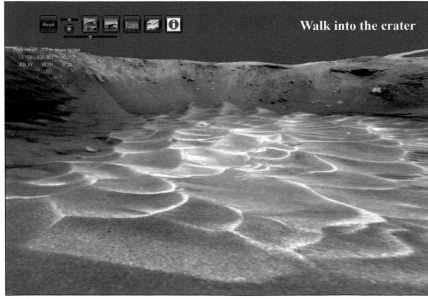

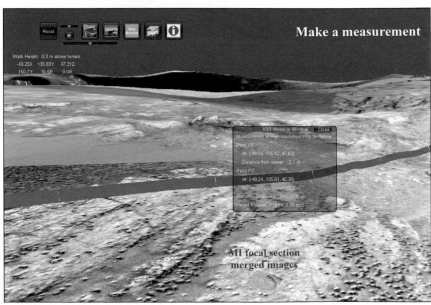

Geologic Map of Big Bend National Park

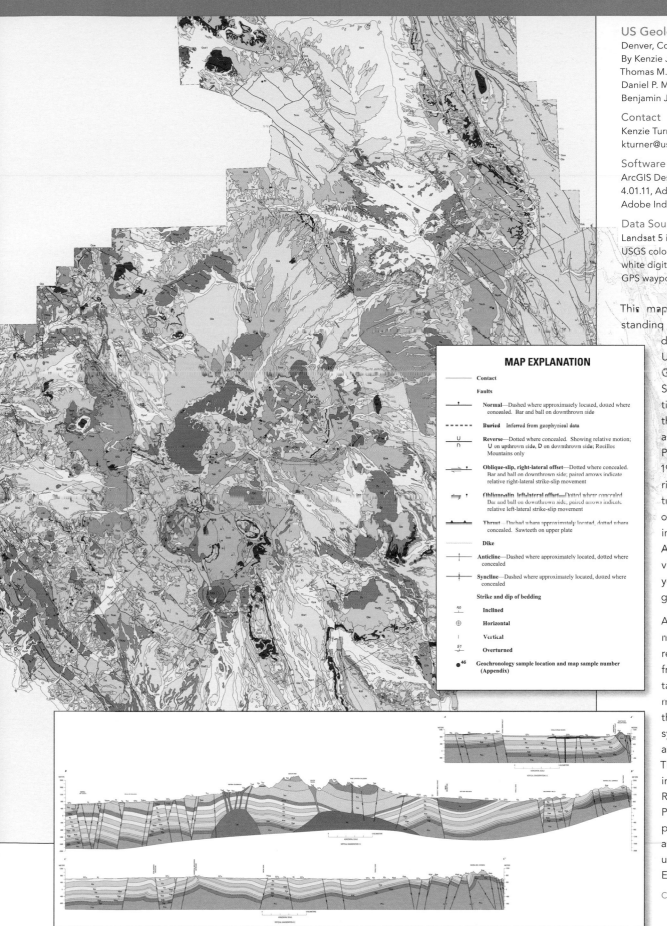

US Geological Survey
Denver, Colorado, USA

By Kenzie J. Turner, Margaret E. Berry, William R. Page, Thomas M. Lehman, Robert G. Bohannon, Robert B. Scott, Daniel P. Miggins, James R. Budahn, Roger W. Cooper, Benjamin J. Drenth, Eric D. Anderson, and Van S. Williams

Contact
Kenzie Turner
kturner@usgs.gov

Software
ArcGIS Desktop 9.2 and 9.3, Cardinal Systems VRONE 4.01.11, Adobe Illustrator CS5, Adobe Photoshop CS5, Adobe Indesign CS5

Data Sources
Landsat 5 imagery; reduced-to-pole aeromagnetic data; USGS color infrared digital orthophotos; USGS black-and-white digital orthophotos; 30-meter digital elevation model GPS waypoints; field mapping

This map portraying the modern geologic understanding of Big Bend National Park (BBNP) was produced through a cooperative effort of the US Geological Survey National Cooperative Geologic Mapping Program, National Park Service, and numerous academic institutions. The purpose of this map is to provide the National Park Service and the public with an updated digital geologic map of BBNP. Previous parkwide mapping was published in 1967 on a geographically distorted planimetric base that lacked topography, which drastically limited GIS based analysis because of an inability to georeference the geologic interpretations depicted on the older map. Additionally, numerous studies since the previous map was published more than forty years ago have significantly advanced the geologic understanding of the park.

ArcGIS was used both for generating new mapping based on field investigation and remotely sensed data and for compiling data from analog and nongeoreferenced digital data sources. The new digital geologic map will aid park managers, researchers, and the public with geologic data analysis, ecosystems management, monitoring, resource assessment, education, and recreation uses. The map depicts recent geologic advances, including a history of the integration of the Rio Grande drainage, a refined Tertiary Period volcanic and intrusive history supported by geochronologic and geochemical analyses, and a comprehensive stratigraphic understanding of Upper Cretaceous to Eocene sedimentary rocks.

Courtesy of US Geological Survey.

MAP EXPLANATION

—— Contact

Faults

Normal—Dashed where approximately located, dotted where concealed. Bar and ball on downthrown side

- - - - **Buried**   Inferred from geophysical data

Reverse—Dotted where concealed. Showing relative motion; U on upthrown side, D on downthrown side; Rosillos Mountains only

Oblique-slip, right-lateral offset—Dotted where concealed. Bar and ball on downthrown side; paired arrows indicate relative right-lateral strike-slip movement

Oblique-slip, left-lateral offset—Dotted where concealed. Bar and ball on downthrown side; paired arrows indicate relative left-lateral strike-slip movement

Thrust—Dashed where approximately located, dotted where concealed. Sawteeth on upper plate

—— **Dike**

Anticline—Dashed where approximately located, dotted where concealed

Syncline—Dashed where approximately located, dotted where concealed

Strike and dip of bedding

50 **Inclined**

⊕ **Horizontal**

| **Vertical**

51 **Overturned**

● 46 **Geochronology sample location and map sample number (Appendix)**

World Shale Resources Map, 2012 Edition

Platts
Westminster, Colorado, USA
By Claude Frank and Erin LeFevre

Contact
Claude Frank
maps@platts.com

Software
ArcGIS Desktop 9.3, Adobe Illustrator CS 5.5, Avenza
MAPublisher, XTools

Data Sources
Platts, Energy Information Administration, US Geological
Survey, Bentek

In recent years, shale has taken the spotlight as the fuel source likely to dominate energy production and consumption in the future. With its global distribution, shale has the potential to transform international energy trading dynamics and even alter the global political and economic landscapes.

Platts' new World Shale Resources 2012 edition wall map presents the primary resources of the emerging global shale industry in striking detail and vivid color. Nearly eighty shale plays in North America, Australia, India, Europe, and Central Asia are shown as well as more than 130 shale basins around the world.

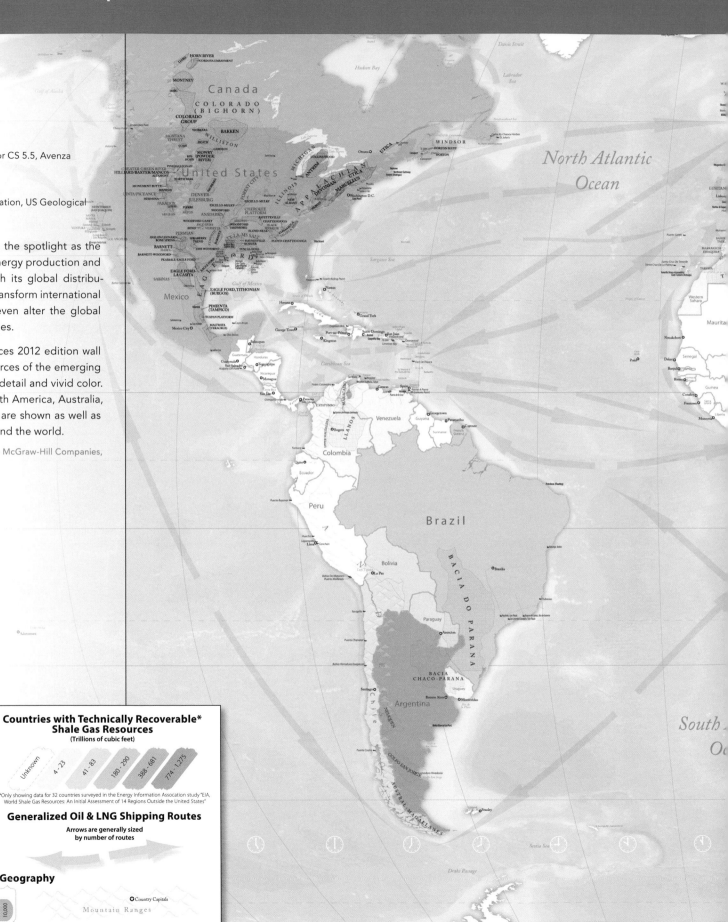

Legend

Shale Basins & Plays

SHALE PLAYS

SHALE BASINS

Other Features

- Major Oil Ports
- LNG Terminals
- Oil Pipelines
- Major Oil Refineries*

*Only refineries with capacity >=200,000BPD are shown

Countries with Technically Recoverable* Shale Gas Resources
(Trillions of cubic feet)

Unknown · 4 - 23 · 41 - 83 · 180 - 290 · 388 - 681 · 774 - 1,275

*Only showing data for 32 countries surveyed in the Energy Information Assocation study "EIA, World Shale Gas Resources: An Initial Assessment of 14 Regions Outside the United States"

Generalized Oil & LNG Shipping Routes

Arrows are generally sized by number of routes

Geography

Ocean Depth (meters)
0 200 1,000 2,000 3,000 4,000 5,000 6,000 7,000 8,000 9,000 10,000

○ Country Capitals

Mountain Ranges

Cartographers: Claude Frank & Erin LeFevre · Data Sources: U.S. Energy Information Administration, Platts Cartography, Bentek GIS Data
For more information about Platts maps visit www.platts.com/mapsandgeospatial. For GIS data and other geospatial solutions visit www.bentekenergy.com.
© 2012 Platts, a Division of the McGraw-Hill Companies, Inc. All rights reserved.

A Geodetic Strain Rate Model for the Pacific-North American Plate Boundary, Western United States

Nevada Bureau of Mines and Geology
Reno, Nevada, USA
By Corné Kreemer, William C. Hammond, Geoffrey Blewitt,
Austin A. Holland, and Richard A. Bennett (authors),
Jennifer Mauldin (cartographer), Daphne D. LaPointe and
Jonathan Price (editors)

Contact
Jennifer Mauldin
mauldin@unr.edu

Software
ArcGIS for Desktop, Adobe Illustrator, geodetic software

Data Sources
Nevada Geodetic Laboratory GPS devices, Nevada Bureau
of Mines and Geology, US Geological Survey

This map presents a model of crustal strain rates derived from GPS measurements of horizontal station velocities. The model indicates the spatial distribution of deformation rates within the Pacific-North America plate boundary from the San Andreas Fault system in the west to the Basin and Range province in the east. As these strain rates are derived from data spanning the last two decades, the model reflects a best estimate of present-day deformation. However, because rapid transient effects associated with earthquakes (i.e., postseismic deformation resulting in curvature of the GPS time series) have been removed from the GPS data, these strain rates can be considered representative of the long-term, steady-state deformation associated with the accumulation of strain along faults.

This model is useful for both seismic hazard and geodynamic studies to understand the activity rates of (known and unknown) faults and the plate tectonic boundary and buoyancy forces that cause the deformation, respectively. In more slowly deforming areas, we expect fewer, smaller earthquakes, and infrequent large earthquakes will have a much longer recurrence time compared to those in rapidly deforming areas.

Courtesy of Nevada Bureau of Mines and Geology.

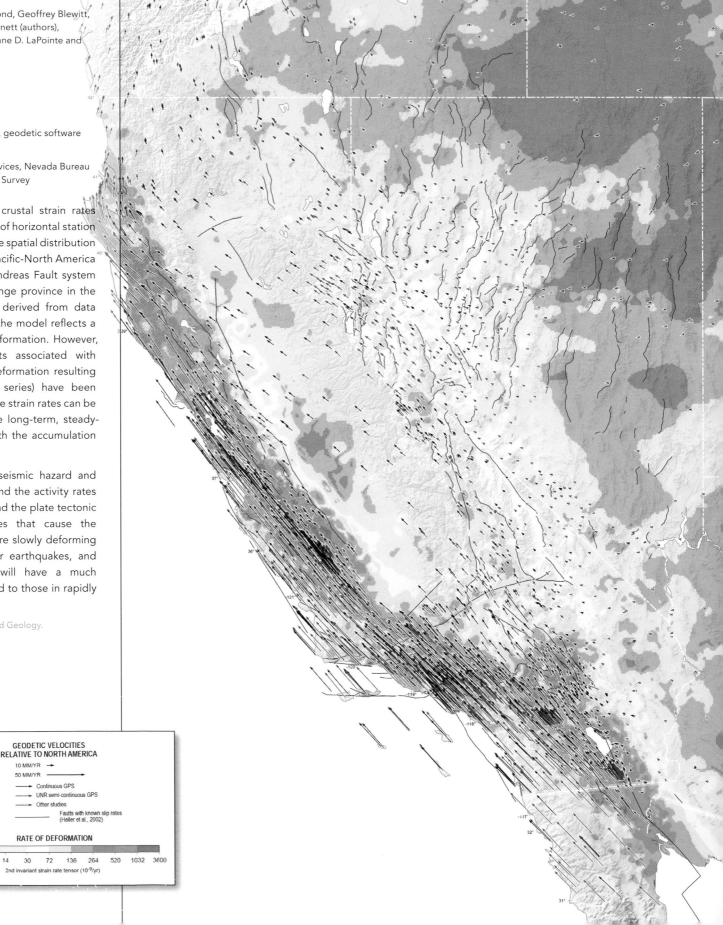

GEODETIC VELOCITIES
RELATIVE TO NORTH AMERICA

10 MM/YR →
50 MM/YR ——→

——→ Continuous GPS
——→ UNR semi-continuous GPS
——→ Other studies

———— Faults with known slip rates
(Haller et al., 2002)

RATE OF DEFORMATION

0   2   6   14   30   72   136   264   520   1032   3600
2nd invariant strain rate tensor (10⁻⁹/yr)

Application of 3D GIS in Planning of Drilling Works

Czech Geological Survey
Prague 1, Prague, Czech Republic
By David Čížek and Jan Franek

Contact
David Čížek
david.cizek@geology.cz

Software
ArcGIS 10 for Desktop

Data Sources
Czech Geological Survey, ISATech, ARCADIS Geotechnika,
Institute of Rock Structure and Mechanics (Academy of
Sciences of the Czech Republic), TU Liberec

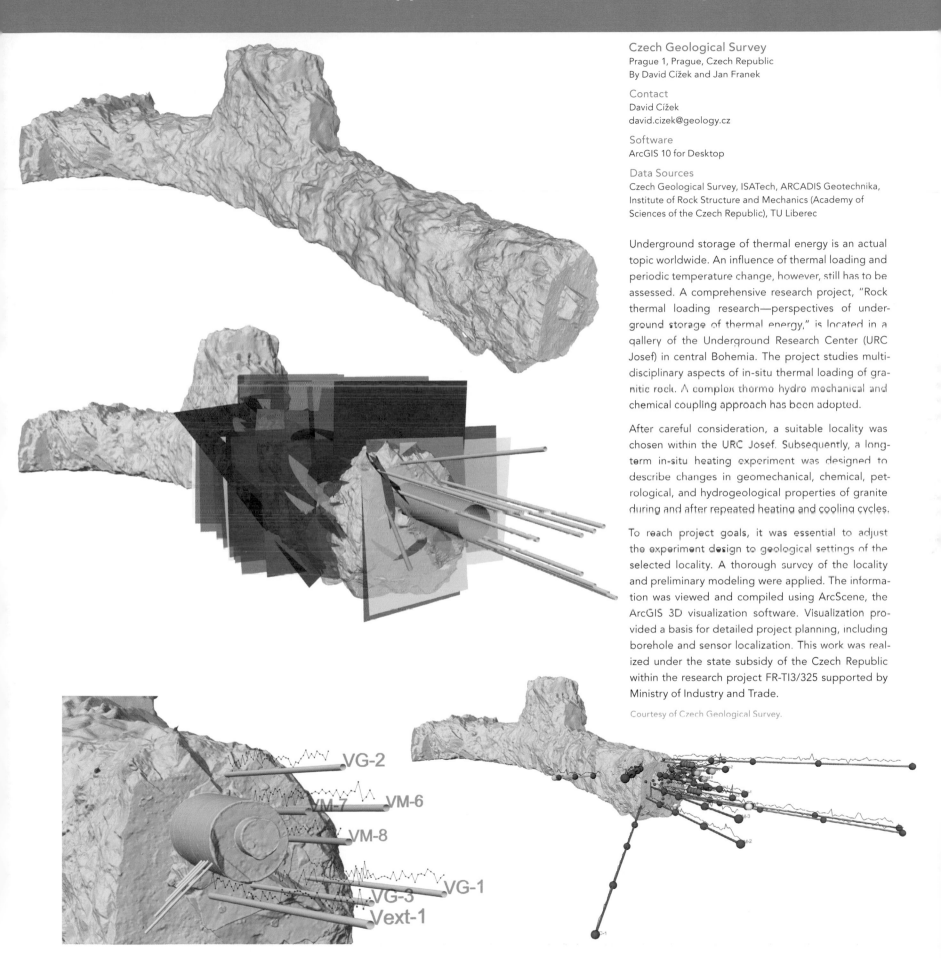

Underground storage of thermal energy is an actual topic worldwide. An influence of thermal loading and periodic temperature change, however, still has to be assessed. A comprehensive research project, "Rock thermal loading research—perspectives of underground storage of thermal energy," is located in a gallery of the Underground Research Center (URC Josef) in central Bohemia. The project studies multidisciplinary aspects of in-situ thermal loading of granitic rock. A complex thermo hydro mechanical and chemical coupling approach has been adopted.

After careful consideration, a suitable locality was chosen within the URC Josef. Subsequently, a long-term in-situ heating experiment was designed to describe changes in geomechanical, chemical, petrological, and hydrogeological properties of granite during and after repeated heating and cooling cycles.

To reach project goals, it was essential to adjust the experiment design to geological settings of the selected locality. A thorough survey of the locality and preliminary modeling were applied. The information was viewed and compiled using ArcScene, the ArcGIS 3D visualization software. Visualization provided a basis for detailed project planning, including borehole and sensor localization. This work was realized under the state subsidy of the Czech Republic within the research project FR-TI3/325 supported by Ministry of Industry and Trade.

Courtesy of Czech Geological Survey.

VG-2
VM-7   VM-6
VM-8
VG-3   VG-1
Vext-1

Greenland: Under the Ice

Community College of Baltimore County
Catonsville, Maryland, USA
By Navneet Sushon

Contact
Scott Jeffrey
sjeffrey@ccbcmd.edu

Software
ArcGIS 10 for Desktop

Data Sources
National Snow and Ice Data Center and Distributed Active
Archive Center at the University of Colorado at Boulder;
Remote Sensing Laboratory of the Byrd Polar Research
Center at the Ohio State University

The country of Greenland is the largest island in the
world. Greenland is not eponymous with its name, as
about 81 percent of the surface area is covered by ice
sheets up to 3.3 kilometers thick. These ice sheets
as a whole form the Greenland ice cap. Covering
an area of approximately 1.8 million square kilometers and with a volume of 2.9 million cubic kilometers,
the Greenland ice cap is second in size only to that
of Antarctica. Although it appears to be frozen solid,
the Greenland ice cap is actually in constant motion,
being replenished by snowfall and flowing steadily
from the center of the island to the coast. The sheer
weight of the ice has depressed the central part of
the island to below sea level, while coastal areas are
steeply elevated.

As with other glacial areas around the world, the
Greenland ice cap is losing its ice mass due to global
climate change. If the Greenland ice cap melts com-
pletely, it is estimated that the world's sea level would
rise by more than seven meters. The purpose of this
project was to visualize and analyze the surface and
bedrock elevations of Greenland using the ArcGIS 3D
Analyst and ArcGIS Spatial Analyst tools. Additionally,
a possible Greenland of the future, a land mass
devoid of any ice, is represented.

Courtesy of Community College of Baltimore County Geospatial
Applications Program.

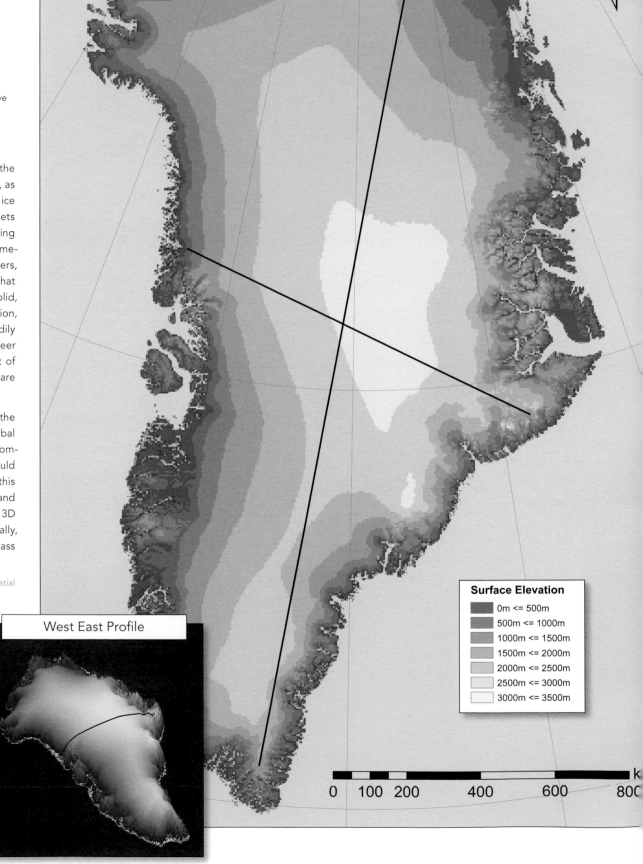

Surface Elevation

- 0m <= 500m
- 500m <= 1000m
- 1000m <= 1500m
- 1500m <= 2000m
- 2000m <= 2500m
- 2500m <= 3000m
- 3000m <= 3500m

k
0   100 200   400   600   800

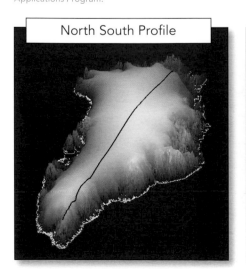

North South Profile

West East Profile

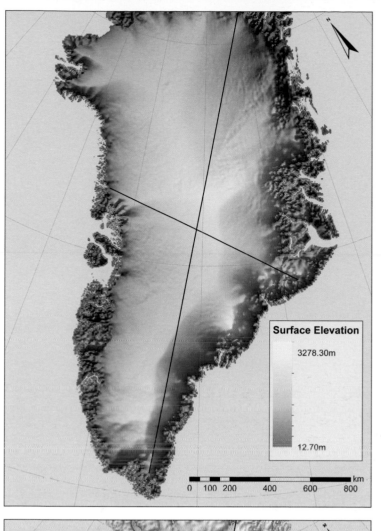

Surface Elevation

3278.30m

12.70m

km
0 100 200   400   600   800

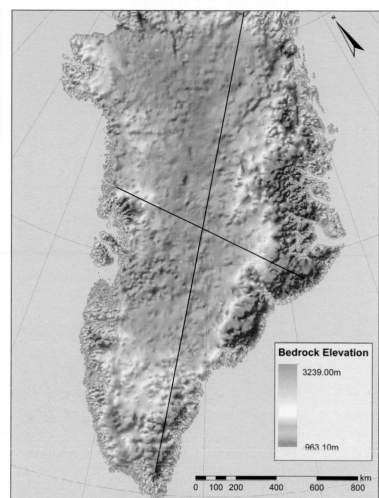

Bedrock Elevation

3239.00m

963.10m

km
0 100 200   400   600   800

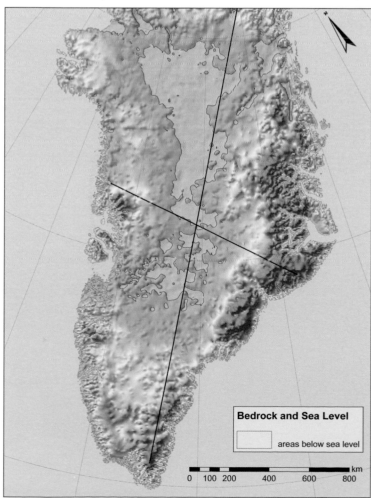

Bedrock and Sea Level

areas below sea level

km
0 100 200   400   600   800

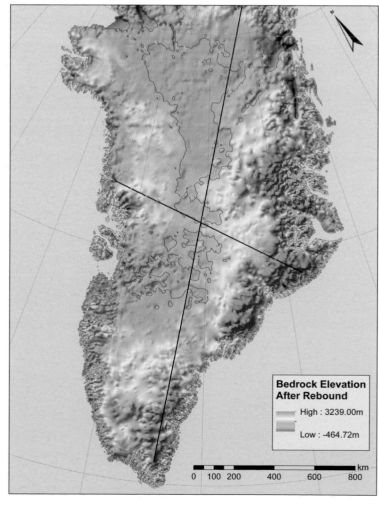

Bedrock Elevation After Rebound

High : 3239.00m

Low : -464.72m

km
0 100 200   400   600   800

Geothermal Favorability Mapping of the Northwest Territories

EBA, a Tetra Tech Company
Vancouver, British Columbia, Canada
By Morgan Zondervan

Contact
Morgan Zondervan
mzondervan@eba.ca

Software
ArcGIS Desktop 9.3, ArcGIS Spatial Analyst

Data Sources
Northwest Territories Geoscience Office; Geological Survey
of Canada Open File Report No. 4173; Geological Survey of
Canada Open File Report No. 6208; Geological Survey of
Canada Open File 3626; Crandall, J. T., and T. L. Sadlier-
Brown (1976); Energy, Mines and Resources

EBA, a Tetra Tech company, was retained by the
Government of the Northwest Territories, Department
of Environment and Natural Resources, to assess the
geothermal potential and to produce a geothermal favorability map of the Northwest Territories
(NWT). The project presented a compilation of existing information and data with respect to the geothermal potential in the NWT and the interpretation of
this data to assess geothermal development potential. Favorability was categorized into five classes
(low, medium-low, medium, medium-high, and high)
based on regional geology, proximity to hot springs,
geothermal gradient, and proximity to and magnitude of seismic events.

Courtesy of EBA, a Tetra Tech Company, and Northwest
Territories Department of Environment and Natural Resources.

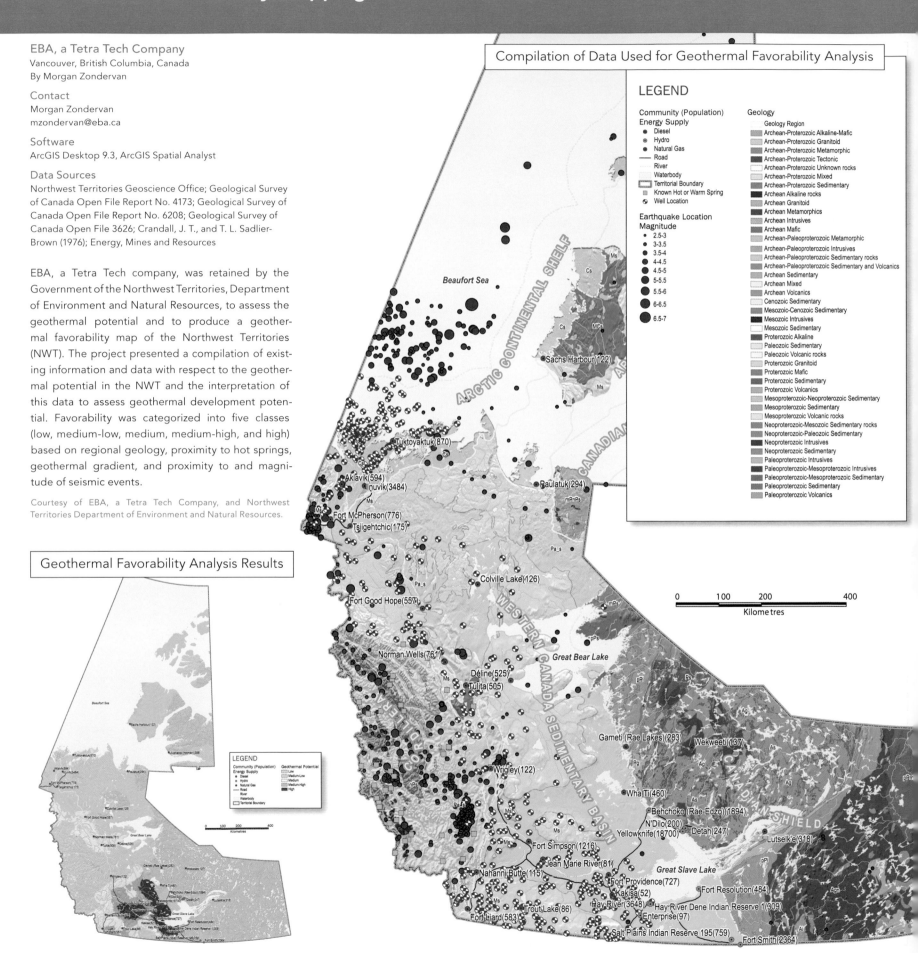

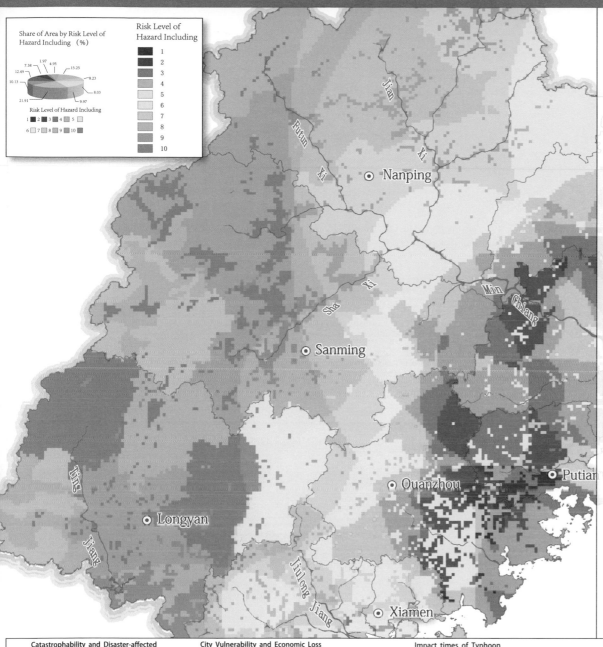

Spatial Information Research Center of Fujian, Fuzhou University
Fuzhou, Fujian, China
By Daichao Li and Jun Luo

Contact
Daichao Li
150909656@qq.com

Software
ArcGIS for Desktop

Data Source
Nature database of Fujian

The Spatial Information Research Center of Fujian focuses its research on a multidisciplinary integration of geoscience, computer science, and communication technologies. It has undertaken many projects researching disaster prevention and mitigation.

Fujian Province suffers from the most severe typhoons in China every year. The map details the present risk of typhoon in Fujian, showing the disaster-affected population, economic loss, rainfall amount, and impact times of typhoons. Also evaluated were the level of catastrophe potential, city vulnerability, and risk hazards.

Courtesy of Daichao Li.

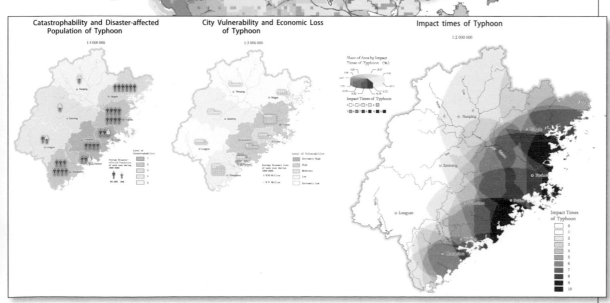

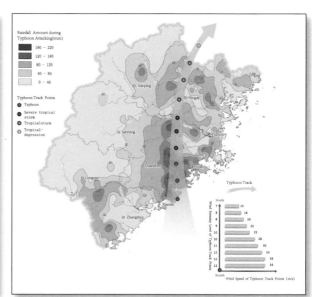

GIS-Based Regional Watershed Improvement Decision Support

AECOM
Washington, DC, USA
By Andy Wohlsperger, Erik Nerrie, and Gray Minton

Contact
Andy Wohlsperger
andy.wohlsperger@aecom.com

Software
ArcGIS 10 for Desktop

Data Sources
Various local, state, and federal agencies

On this collaborative federal/state/local partnership, an innovative GIS-based decision support tool was developed to perform rapid watershed assessments and identify areas in greatest need for watershed improvements. This partnership, led by the US Army Corps of Engineers, provides design and construction assistance to nonfederal interests in southeastern Pennsylvania for water-related environmental infrastructure and resource protection projects. Working inside a highly fragmented political framework, a major challenge of the project was to identify regional issues of highest concern and to develop consensus on locations of regional priority for integrated watershed improvements. Numerous participatory exercises were incorporated into the project to build support among stakeholders and to provide transparency in the watershed assessment and prioritization process.

To conduct the watershed assessments and prioritizations, a custom, easy-to-use GIS extension was developed to synthesize available watershed information and prioritize areas for watershed improvements based on a set of user-defined prioritization weighting factors. Stakeholders are able to conduct their own prioritizations, using custom inputs and user-defined weighting factors, to better understand the

prioritization process and to support local watershed planning efforts. This customizability allows the decision support tool to be used at both the federal and local levels, supporting integrated watershed planning at all geographic scales.

Courtesy of AECOM Technology Corp. Inc.

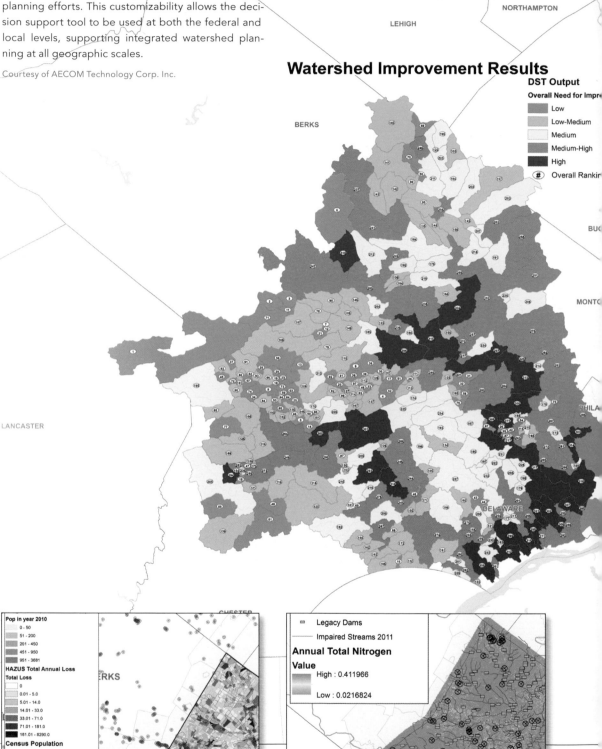

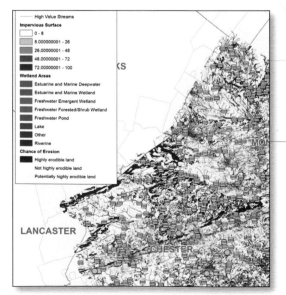

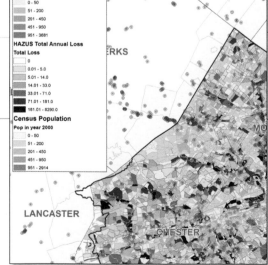

Bathymetry and Volume—Swimming River Reservoir

New Jersey American Water
Voorhees, New Jersey, USA
By Christopher Kahn

Contact
Christopher Kahn
christopher.kahn@amwater.com

Software
ArcGIS for Desktop

Data Sources
New Jersey American Water, New Jersey Geographic
Information Network, Monmouth County (New Jersey) GIS,
US Geological Survey

New Jersey American Water is reviewing water storage capacity in one of its systems as part of long-term capacity planning. The engineering and production groups wanted a bathymetry map that could show how sedimentation was affecting capacity, as well as display the present day maximum volume. Engineers also requested that the map explain how potential changes in the elevation of the water level, above and below the current dam height, might affect reservoir volume.

Over 1,500 depth soundings were recorded with sonar and GPS. When creating a digital elevation model (DEM) from points alone, GIS interpolation methods will incorrectly smooth over details of topography, such as riverbeds and floodplain boundaries. To ensure the analysis honored submerged topography, detailed layers of control contours, derived from historical and field-gathered information, were created. Once these were combined with the depth points, the GIS created a highly accurate DEM. Contours were derived from the DEM. Techniques known as the Swiss style of hillshading and hypsometric tinting help the final map appear to have depth.

Several GIS analyses were performed with the new data. The cut/fill analysis compared a 1986 model to the 2011 model and provided insight on how sedimentation has affected the water body over twenty-five years. The new DEM was also used to calculate actual and potential volumes at 1-foot intervals. Knowing this information helps operators make real-time decisions about production, as well as helps engineers plan for growth.

Courtesy of New Jersey American Water.

Urban Connectivity: Spatial Connections in Portland's Central City

City of Portland Bureau of Planning and Sustainability
Portland, Oregon, USA
By Derek W. Miller

Contact
Derek Miller
derek.miller@portlandoregon.gov

Software
ArcGIS 10 and 10.1 for Desktop, SUTD/MIT City Form Lab—
Urban Network Analysis Toolbox

Data Source
City of Portland

A city's urban form influences interaction between a city's population and the urban environment. Understanding the relationship between the city's population and the natural and built environments is a key piece of the urban design and planning process.

The map highlights the spatial connectivity of Portland's urban core. The goal of this analysis is to identify those areas of the city that should experience higher levels of pedestrian traffic based on the built infrastructure. This is accomplished in ArcGIS by using SUTD/MIT City Form Lab's Urban Network Analysis Toolbox to measure the betweenness of pedestrian-accessible network intersections.

The image illustrates the role of the Willamette River, which dissects the center of the urban core, placing a significant bias on the bridgeheads, the streets that provide access to bridges, and locations that act as collection points between the outer and inner urban core.

Courtesy of City of Portland Bureau of Planning and Sustainability.

connectivity as measured by betweenness
high connectivity

low connectivity

Best Path Planning for Lunar Rover on Multicrater Lunar Surface

Peking University
Beijing, China
By Huiling Wang

Contact
Huiling Wang
wanghuiling2005@126.com

Software
ArcGIS for Desktop, ArcGIS Spatial Analyst

Data Sources
US Planetary Data System, Lunar Reconnaissance Orbiter Camera (Lunar Orbitar Laser Altimeter/Lunar Reconnaissance Orbiter Radio Science Planetary Data System Data Node)

This series of maps was created at the School of Earth and Space Sciences of Peking University as a student project, primarily using ArcGIS Spatial Analyst tools. The maps show macro-accurate path planning for the lunar rover when it travels on the multicrater lunar surface. Figure A is the rendering map of the entire lunar digital elevation model data and marked in accordance with the international astronomical organization naming of the lunar crater. Figure B extracts a scope ranging 50E–70E, 0S–15S from figure A, with the path planned by spatial analysis tools. Figure C selects the optical data of the area surrounding the hit point of the ChangE-1 and performs path planning with elevated precision.

Courtesy of Huiling Wang.

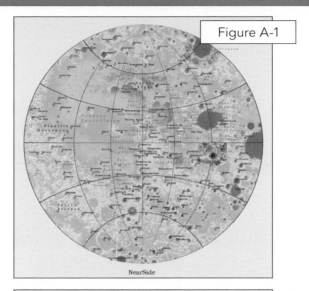

Figure A-1

NearSide

Figure A-2

FarSide

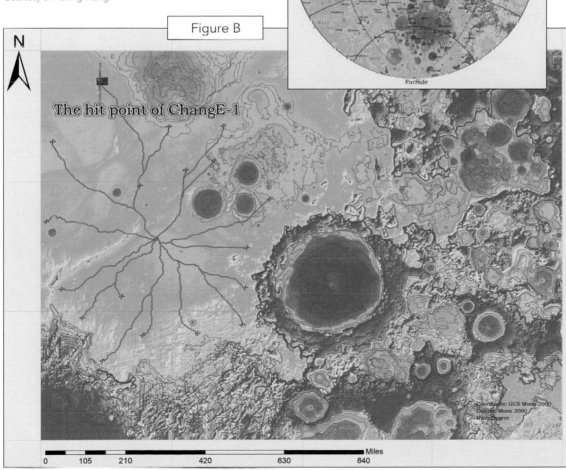

Figure B

N

The hit point of ChangE-1

Coordinate: GCS Moon 2000
Datum: Moon 2000
Units: Degree

Miles
0   105   210   420   630   840

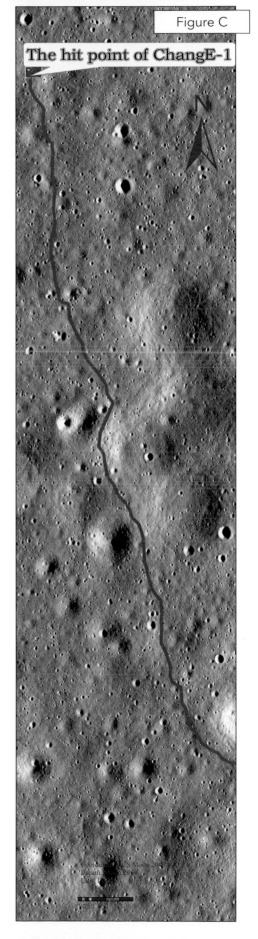

Figure C

The hit point of ChangE-1

Land-Use Planning and Urban Development Projects

L'Institut d'aménagement et d'urbanisme de
la Région Île-de-France (Urban Planning and
Development Agency for the Paris Region)
Paris, Île-de-France, France
By Xavier Opigez

Contact
Xavier Opigez
xavier.opigez@iau-idf.fr

Software
ArcGIS 10.1 for Desktop, CityEngine 2010.2,
SketchUp Pro 6, Lumion 2.2.1

Data Sources
InterAtlas (digital orthophoto 2008), BD TOPO (building
footprints) IGN

L'Institut d'aménagement et d'urbanisme de la
Région Île-de-France (IAU îdF) chose to develop a 3D
model to deal with new planning challenges such as
urban intensification, sustainable development, and
physical integration of new buildings. 3D modeling is
becoming a tool to reveal and analyze the urban den-
sification potential and to visualize proposed development projects. 3D is very useful to simulate the
impact of natural or industrial risks, such as flooding
or gas leaks on communities and businesses whether
existing or proposed..

It is also useful to visualize the impact of high-rise
construction projects as shown here. This image
was produced as part of a series of workshops on
the evolution of the Paris metropolitan landscape.
The skyscraper in the foreground and the two blue
skyscrapers in the background are proposed urban
developments for the Paris central business district
(La Défense). 3D modeling of this area was used to
analyze the visual impact on the skyline of these three
high-rise buildings.

The 3D model was built with CityEngine software
generating over 25,000 buildings by a procedural
mode. Development projects and geospecific buildings were built with SketchUp.

Radio Tower Height Signal Strength Analysis

San Bernardino County Information Services
Department
San Bernardino, California, USA
By Brent Rolf

Contact
Mike Cohen
mcohen@isd.sbcounty.gov

Software
ArcGIS 10 for Desktop, ComStudy 2.2

Data Source
San Bernardino County ISD (street network, city limits,
airports); San Bernardino County ISD Radio Division (signal
strength/propagation map layers); US Geological Survey
hillshade, place-names

In October 2011, San Bernardino County needed to augment its radio coverage in the Victor Valley region. The county had recently built a new government center in this region, and it was advantageous to place a new radio tower on the center's site. To provide the necessary radio coverage, a substantial tower had to be built. To support the county's public outreach program regarding the need for the tower, the Radio Division sought help from the Information Services Department's Geographical Information Management Services (GIMS) Team to prepare map products depicting the areas covered by the new radio tower at varying heights.

The county's communication specialists were able to produce graphic files, using the ComStudy 2.2 software, showing signal strength and coverage by tower height. The GIMS Team was able to georeference the graphic files and produce maps showing signal strength and coverage for the four tower scenarios. The ability of the maps to provide a visual representation of how radio coverage varied with tower height allowed the county's decision makers to choose a tower height that provided sufficient coverage and minimized cost and aesthetic concerns. The outcome of this analysis was the choice of the 175-foot tower as the optimum solution.

Courtesy of San Bernardino County.

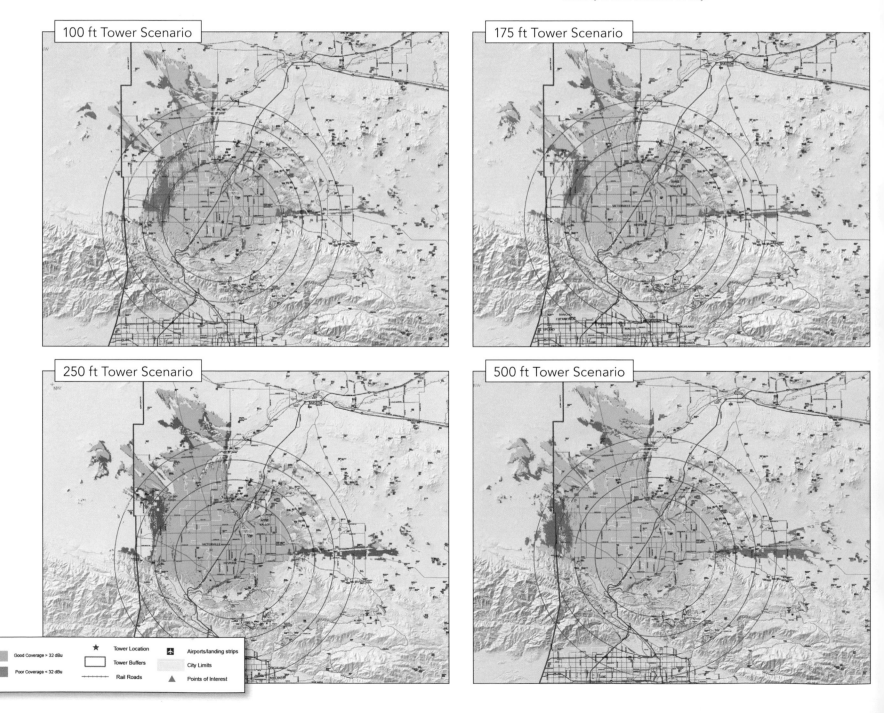

100 ft Tower Scenario

175 ft Tower Scenario

250 ft Tower Scenario

500 ft Tower Scenario

Good Coverage > 32 dBu
Poor Coverage < 32 dBu
Tower Location
Tower Buffers
Rail Roads
Airports/landing strips
City Limits
Points of Interest

KTU+A

San Diego, California, USA
By Joe Punsalan

Contact

Joe Punsalan
joe@ktua.com

Software

ArcGIS 10 for Desktop, Adobe InDesign

Data Sources

San Diego Association of Governments, US Census,
Statewide Integrated Traffic Research System

The San Diego County Walkability Index was a collaborative effort between WalkSanDiego and KTU+A to assess the region's efforts in improving pedestrian infrastructure. This model was developed to determine locations throughout the county where the highest propensity of walking occurred. The areas with higher propensity were then imported into a custom mobile phone application to allow volunteers to field check these streets for amenities and deficiencies. The data collected was then scored along with other policies, programs, and projects that have already been or are in the process of being developed to improve pedestrian access within each city. Scores were tabulated to identify the cities that showed the most progress in improving walkability.

The model was composed of three pedestrian submodels: Attractors, Generators, and Barriers.

Attractors consisted of parks, schools, transit stops, retail, public facilities, and major attractions. Walk-time polygons were generated using ArcGIS Network Analyst from each attractor, and higher scores were given to the polygons closest to the attractor. Generators were made up of existing and future population and employment densities, commuting demographics, vehicle ownership, and income. Barriers consisted of pedestrian collisions, speed, number of travel lanes, traffic volumes, and topography.

KTU+A is a planning and landscape architecture firm that specializes in active transportation planning (bicycle, pedestrian, transit, Americans with Disabilities Act, and trail), urban, and federal planning. KTU+A also specializes in 3D modeling, simulations, and landscape architecture and design.

Courtesy of KTU+A.

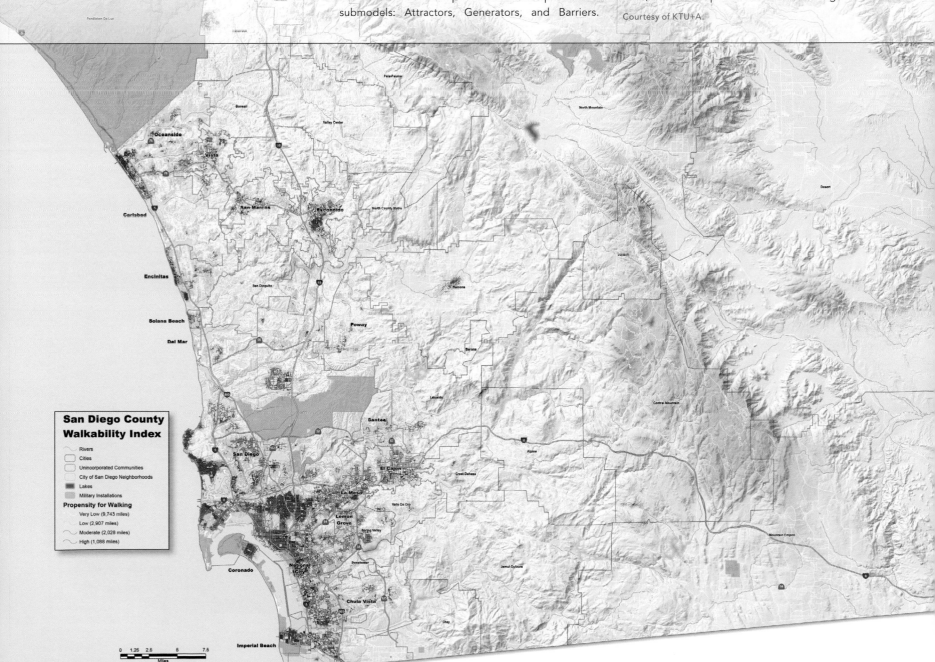

San Diego County Walkability Index

- Rivers
- Cities
- Unincorporated Communities
- City of San Diego Neighborhoods
- Lakes
- Military Installations

Propensity for Walking

- Very Low (9,743 miles)
- Low (2,907 miles)
- Moderate (2,028 miles)
- High (1,088 miles)

0 1.25 2.5   5   7.5
Miles

Automatic Bus Routes Schematic Map

Adalya Financial Consulting & Systematics Technologies
Tel Aviv, Israel
By Mor Yaffe and Talya Waserman

Contact
Mor Yaffe
mory@systematics.co.il

Software
ArcGIS 10 for Desktop, ArcGIS Schematics, Maplex

Data Sources
Adalya bus routes layers, Survey of Israel streets layer

Adalya Financial Consulting & Systematics Technologies were asked, as part of a pilot, to find a way to automate the production of informative signs for a bus station. These signs are designed for delivering useful information related to bus travel from each station, including a reference map of the station's immediate area, a schematic (orthogonal) map of the bus routes stopping at the station, and a linear diagram of each bus route's stations.

Using ArcGIS Schematics, an automated workflow was developed in which bus routes, bus stations, and street data were integrated into a unified geometric network. Then, ArcGIS Schematics was used to create a schematic diagram aligning the routes and streets to straight angles. Additional edit operations were automated to create the visual separation of the various routes going through the same street segments. The schematic Smart Tree algorithm was used to generate the linear single bus route diagrams.

Automation of the rest of the sign components (large-scale geographic map of the station vicinity, timetables, etc.) was achieved using standard ArcObjects libraries.

The entire automated process is now in pilot phase, after which the actual mass production of the signs would be considered.

Courtesy of Adalya Financial Consulting, Survey of Israel.

City of Lawrence
Lawrence, Kansas, USA
By Micah Seybold, Emily Lubliner, Margretta de Vries, and Derek Meier

Contact
Micah Seybold
mseybold@lawrenceks.org

Software
ArcGIS 10 for Desktop, Adobe InDesign

Data Sources
City of Lawrence, the University of Kansas

The 2012–2013 Lawrence Transit System Map provides transit riders in Lawrence, Kansas, with an overview of bus service across town and on the campus of the University of Kansas. It is one of several map products created annually by the Lawrence Transit Map Team to help bus patrons plan their trips. The team consists of representatives from both the City of Lawrence and the University of Kansas. Design decisions are made by the team as a whole, with an emphasis on making easy-to-read maps and guides.

The city provides enterprise GIS basemap data and cartographic production using ArcGIS for Desktop software to complete the system maps, and university staff prepares the maps for production in separate design software. Lawrence was recognized by the Federal Transit Administration for the largest increase in ridership among urban transit providers in Kansas for 2010 and 2011. It was also the first transit system in Kansas to operate hybrid buses.

Courtesy of City of Lawrence, Kansas.

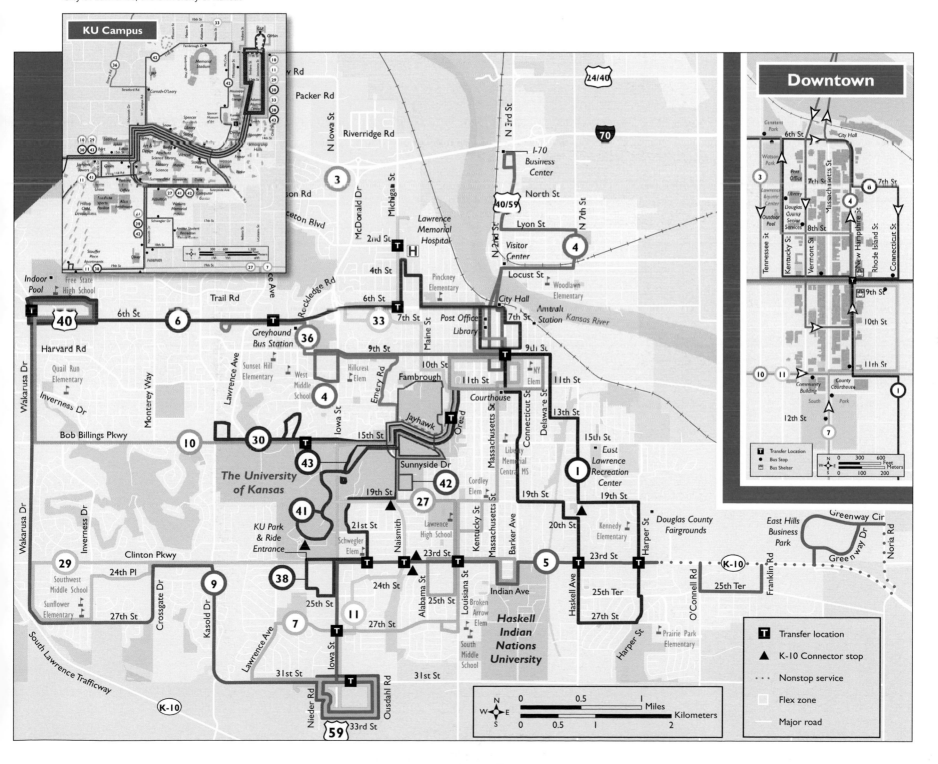

Rosedale Highway Corridor Improvements

Parsons Corp.
San Jose, California, USA
By Eric Coumou

Contact
Eric Coumou
eric.coumou@parsons.com

Software
ArcGIS 10 for Desktop

Data Sources
Thomas Roads Improvement Program, Parsons, City of Bakersfield

A series of maps was generated for the Rosedale Highway corridor in Bakersfield, California. The footprint of the highway project, along with newly designed sidewalks and ramps, is overlaid on aerial photography, along with parcel lines and underground utilities. The maps show planners which parcels need to be acquired (takes) or parcels that are only partially impacted (partial takes). They also show planners which utilities will have to be relocated and help planners mitigate project impacts on the endangered San Joaquin Valley kit fox. The Rosedale Highway corridor is one of many projects for the Thomas Roads Improvement Program in the Bakersfield area.

Courtesy of Parsons Corp.

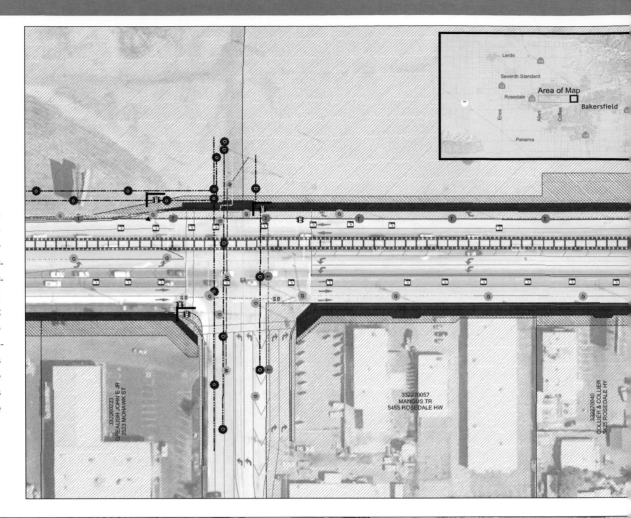

Utility Poles, etc.

↗ Highway Sign	KB Fox dens
Hydrant	○ Manholes
Light	M Monuments
○ Parking Lot Light	Oil Donkey
Railroad Crossing	PhotoPoints
Traffic Signal	Sewer Pipes
Traffic Signal and Light	Partial Takes
Walk Sign	Temporary Construction Easement
	Parcels Effected by Roadway Construction
	Bakersfield Parcels

G — Gas - SCG
SD --- UT-stormd
T -·- UT-teleph-x
TC --- Telecom - AT&T
E --- Electric - PGE
O --- Oil - Chevron
W — Water - Vaughn

TRIP Project Area

Rosedale Highway

58

332020809

CHEVRON USA INC
5401 ROSEDALE HW
332270032

332270016
METAL WORLD INC
5303 ROSEDALE HY

332260025
WANAMAKER
JEFF ET AL

CHAMPLIN FAMILY TR
5401 ROSEDALE HW
332270024

PARKER LN

BNSF RAILWAY CO
332280016

DEED

DOWNES FAMILY
5200 ROSEDALE
332020510

Rosedale Highway

332020817

332260124
BIG WEST OF
CALIFORNIA LLC

332260355
6451 ROSEDALE HY

332260264
GREALISH JOHN E JR
6045 ROSEDALE HW

Rosedale Highway Corridor Improvements

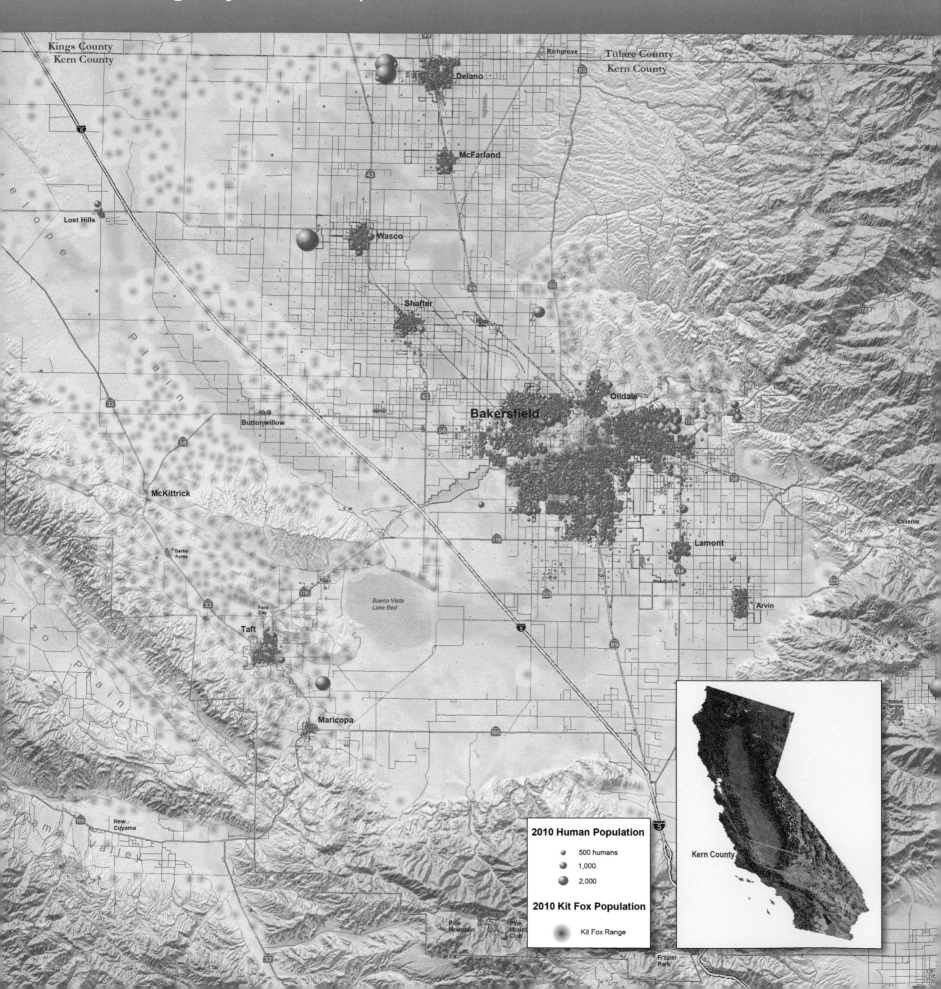

2010 Human Population

- 500 humans
- 1,000
- 2,000

2010 Kit Fox Population

Kit Fox Range

Kern County

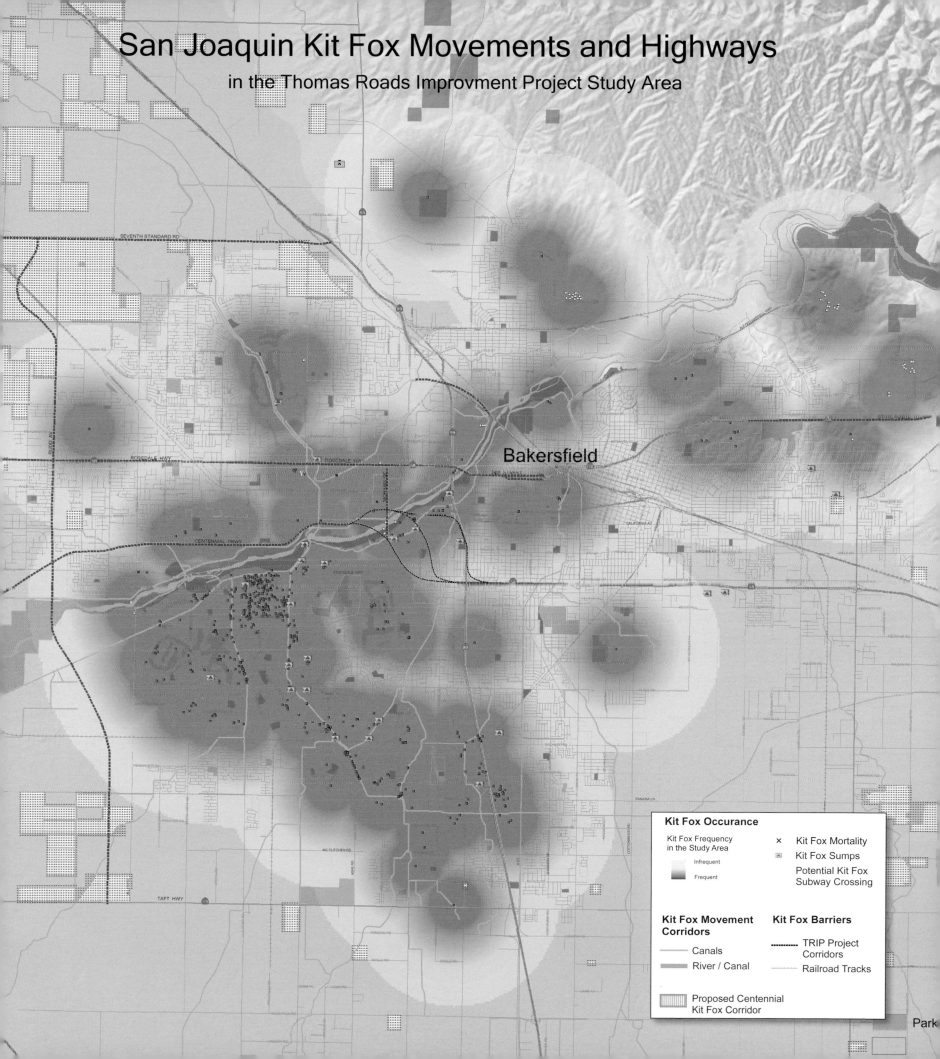

Southern California Passenger Rail

San Diego Association of Governments
San Diego, California, USA
By Danny Veeh (SANDAG) and Victoria Giebel (North County Transit District)

Contact
Danny Veeh
Danny.Veeh@sandag.org

Software
ArcGIS for Desktop, Adobe Illustrator

Data Sources
SANDAG, Caltrans

This map depicts all intercity and commuter passenger rail lines and stations within Southern California from San Luis Obispo to San Diego. A comprehensive map with all rail operators in the region was needed to help passengers visualize the different services and connections that are available. The map was developed by the San Diego Association of Governments (SANDAG) to accompany the Southern California Passenger Rail Timetable which, for the first time, combines the schedules of Amtrak, Metrolink, and COASTER trains, which share tracks and stations.

The map and timetable were developed by the Los Angeles-San Diego-San Luis Obispo (LOSSAN) rail corridor agency, which seeks to increase ridership, revenue, capacity, reliability, and safety on the coastal rail corridor. This map and timetable are a new resource to rail passengers in Southern California that will encourage more ridership and revenue by providing complex train information in an easily understandable format. GIS was applied to develop the basemap, geocode rail lines and stations, and create simplified data layers that maintain geographic accuracy while improving legibility.

Courtesy of San Diego Association of Governments regional information system, 2012.

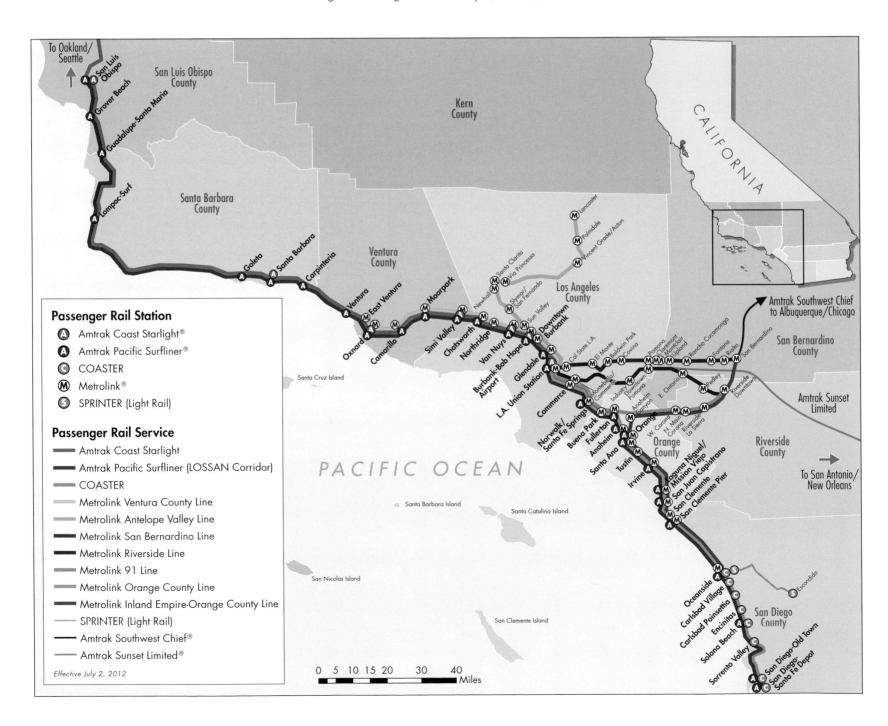

Passenger Rail Station
Ⓐ Amtrak Coast Starlight®
Ⓐ Amtrak Pacific Surfliner®
Ⓒ COASTER
Ⓜ Metrolink®
Ⓢ SPRINTER (Light Rail)

Passenger Rail Service
— Amtrak Coast Starlight
— Amtrak Pacific Surfliner (LOSSAN Corridor)
— COASTER
— Metrolink Ventura County Line
— Metrolink Antelope Valley Line
— Metrolink San Bernardino Line
— Metrolink Riverside Line
— Metrolink 91 Line
— Metrolink Orange County Line
— Metrolink Inland Empire-Orange County Line
— SPRINTER (Light Rail)
— Amtrak Southwest Chief®
— Amtrak Sunset Limited®

Effective July 2, 2012

Mobility Corridor Atlas

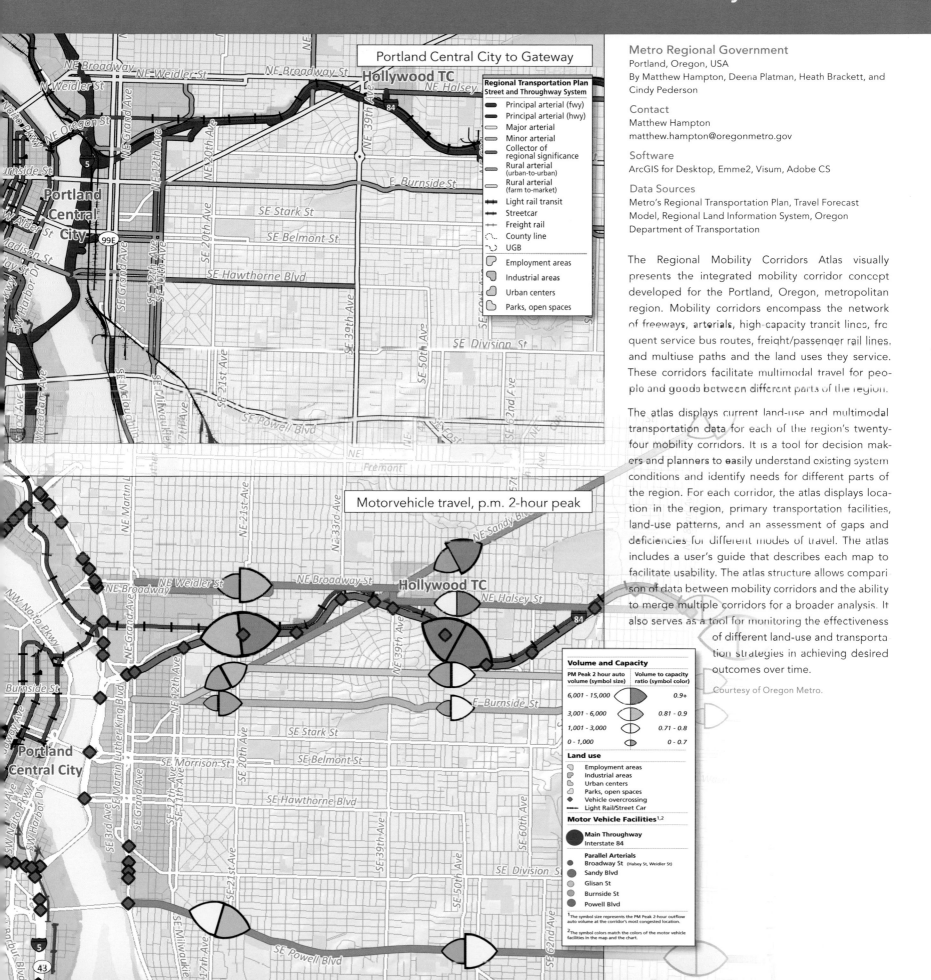

Metro Regional Government

Portland, Oregon, USA
By Matthew Hampton, Deena Platman, Heath Brackett, and Cindy Pederson

Contact

Matthew Hampton
matthew.hampton@oregonmetro.gov

Software

ArcGIS for Desktop, Emme2, Visum, Adobe CS

Data Sources

Metro's Regional Transportation Plan, Travel Forecast Model, Regional Land Information System, Oregon Department of Transportation

The Regional Mobility Corridors Atlas visually presents the integrated mobility corridor concept developed for the Portland, Oregon, metropolitan region. Mobility corridors encompass the network of freeways, arterials, high-capacity transit lines, frequent service bus routes, freight/passenger rail lines, and multiuse paths and the land uses they service. These corridors facilitate multimodal travel for people and goods between different parts of the region.

The atlas displays current land-use and multimodal transportation data for each of the region's twenty-four mobility corridors. It is a tool for decision makers and planners to easily understand existing system conditions and identify needs for different parts of the region. For each corridor, the atlas displays location in the region, primary transportation facilities, land-use patterns, and an assessment of gaps and deficiencies for different modes of travel. The atlas includes a user's guide that describes each map to facilitate usability. The atlas structure allows comparison of data between mobility corridors and the ability to merge multiple corridors for a broader analysis. It also serves as a tool for monitoring the effectiveness of different land-use and transportation strategies in achieving desired outcomes over time.

Courtesy of Oregon Metro.

Expansion and Shrinkage of Transport Space in Russia

Moscow State University, Faculty of Geography
Moscow, Russia
By Pavel Sapanov

Contact
Pavel Sapanov
psapanov@esri-cis.ru

Software
ArcGIS for Desktop

Data Sources
Basemap: OpenStreetMap geometry; revenues taken from official statistics site; other data (ticket prices) taken from public sources

The anisotropy of transport space—its discontinuity— has always been an important issue for Russia and its neighboring countries. The shrinking transport space can be characterized as increased accessibility to the elements within that space. In the recent historical past, the studied countries used to be one state with strong relationships including common sociopolitical, economic, and transport systems. This map was part of a study into the disproportion between passenger demand and availability of rail and air transport and especially changes that occurred during the period 1985–2012.

The two tasks of the research were to describe the results in space and in time and to visualize them. To accomplish the first task, the coefficient of social transport accessibility (taking into account the average cost of travel and income of inhabitants) of a region was calculated and a data preparation algorithm was created. The second task was completed using ArcGIS for Desktop. Maps were created using raster analysis, ArcGIS 3D Analyst, interpolation methods, and appropriate symbology, which displayed the changes in transport accessibility for passengers from different territories in the studied area

The proposed method of analyzing transport accessibility to territories, taking into account the cost of travel and the income of the residents, is viable for studying areas of any scale. For example, it was shown that for people who live in Moscow, the transport space has been shrinking over the studied time period, increasing accessibility to all neighbor regions. During the research, the regions have been grouped according to the dynamics of transport accessibility. The reasons for good or bad dynamics have been explained. The intensity of the development of transport in Russia and neighboring countries correlates with territories with either high revenues or have strong state support.

Courtesy of Pavel Sapanov.

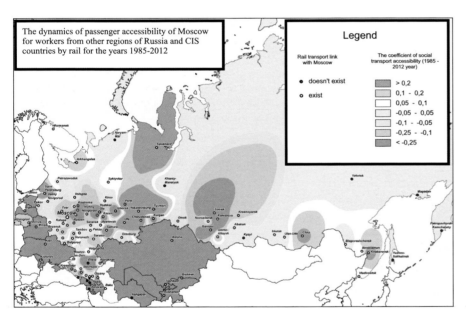

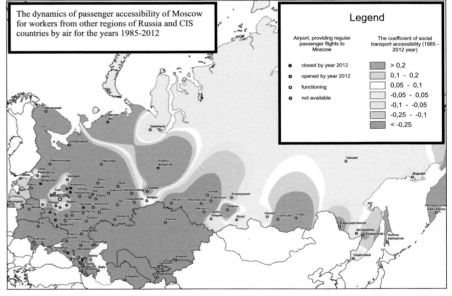

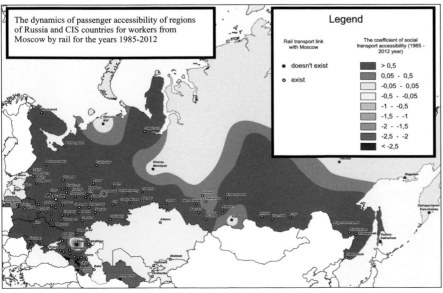

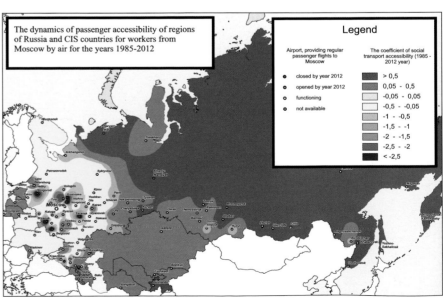

Electromobility in the Bavarian Forest

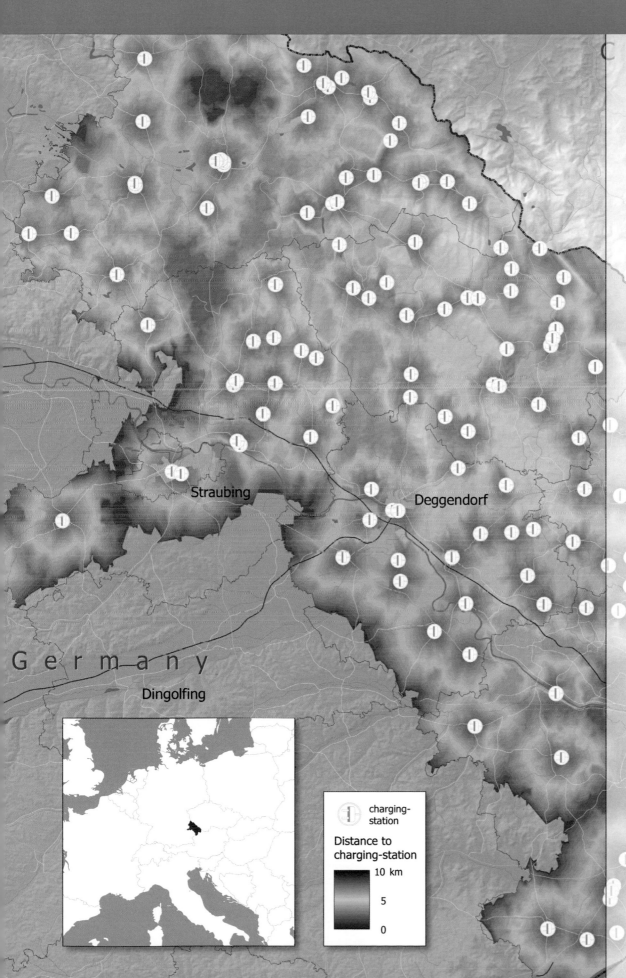

University of Applied Sciences Deggendorf
Freyung, Bavaria, Germany
By Roland Zink and Michael Burghart

Contact
Roland Zink
roland.zink@hdu-deggendorf.de

Software
ArcGIS 10 for Desktop, ArcGIS Network Analyst

Data Sources
Vektor500, Esri Data & Maps 2006, OpenStreetMap, Shuttle
Radar Topography Mission

This map shows the early stages of the infrastructure
planning process where parameters, including opti-
mal charging-station positions, are researched. Each
car in the fleet was equipped with a GPS data logger
and captured additional parameters such as speed,
elevation, bearing, battery condition, temperature,
slope, and driver profiles correlated to the respective
geoposition. The captured data is essential for the
research process and enables precise scientific pre-
dictions on the cruise range depending on parameters like driver behavior in the context of a specific
region. Consequently, these findings can be used
in applications for driver adapted navigation systems that could be implemented within an intelligent
driver-assistant system.

Courtesy of University of Applied Sciences Deggendorf, Project
E-Wald.

Straubing

Deggendorf

G e r m a n y

Dingolfing

Passau

charging-
station

**Distance to
charging-station**

10 km

5

0

Geographic Analysis Utilized in Transmission Routing in the San Juan Basin

Tetra Tech
Denver, Colorado, USA
By Matthew Smith, Jennifer Chester, and Jake Engelman

Contact
Matthew Smith
matthew.smith@tetratech.com

Software
ArcGIS Desktop 9.3

Data Sources
Bureau of Transportation Statistics, Colorado Department of Transportation, Colorado State University, New Mexico Resource Geographic Information System, Colorado Division of Wildlife, US Bureau of Land Management, New Mexico Game and Fish

Identifying potential routes for high-voltage transmission lines is a complex process that requires analyzing large amounts of information to select the most suitable location for the line. The San Juan Basin Energy Connect used GIS to develop a defensible approach to route selection for the proposed 230-kilovolt transmission line from the Farmington area in northwestern New Mexico to Ignacio, Colorado.

The route identification process used GIS throughout the various phases of the project:

In phase 1, preliminary corridors were identified based on the required interconnection points and an opportunity and constraint analysis of the resource data. Available resource data for land use and natural and cultural resources was gathered, mapped, and categorized based on each resource's suitability or sensitivity to construction, operation, and maintenance of a transmission line.

In phase 2, corridors were refined, expanded, or removed based on factors such as public input, identification of sensitive resources, and additional field review. Preliminary routes were then identified within the corridors and a comparative analysis of routes completed based on quantifiable data.

In phase 3, final adjustments were made to the alternative route options, and a preferred route and feasible alternatives were identified. The selected routes were then carried forward for analysis under the National Environmental Policy Act.

The route selection process used GIS for cartographic production of dozens of maps for display of resource data and project alternatives, geoprocessing of project area information for analysis, as well as linking of database outputs of information quantified for comparative analysis purposes.

Courtesy of Tetra Tech, Tri-State Generation and Transmission Association

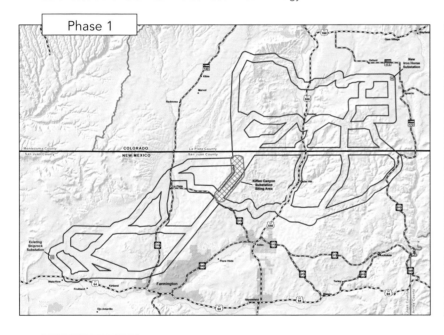

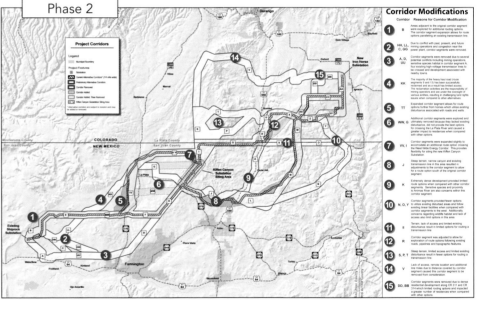

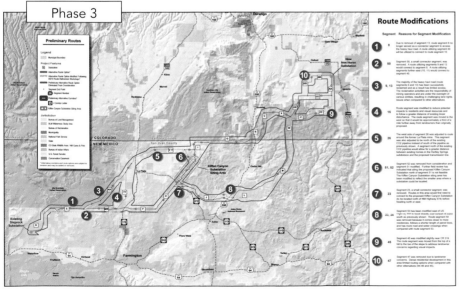

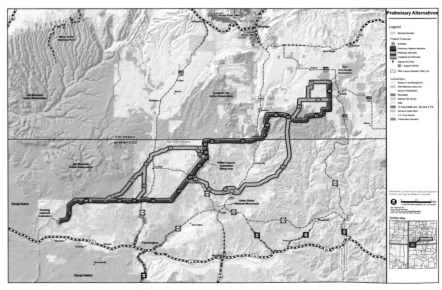

Saudi Arabia Solar Resource

Sempra Energy
San Diego, California, USA
By Joel Griffin

Contact
Joel Griffin
gisgriffin@gmail.com

Software
ArcGIS 10 for Desktop, ArcGIS Spatial Analyst

Data Sources
Solar Reserve, Sempra Energy, Saudi Electric Company, Shuttle Radar Topography Mission, Jet Propulsion Laboratory, World Database on Protected Areas, Food and Agriculture Organization of the United Nations

The Sempra Energy Saudi Arabia Solar Resource map was created to show high-level internal management and outside partners the potential areas for solar energy development. The Kingdom of Saudi Arabia provides an ideal location for solar energy development, with most of the country providing more than 6 kilowatt hours per square meter per day. Sempra Energy wanted to identify and rank potential sites to focus efforts. One major advantage was being able to use any land available because it is all owned by the kingdom. A major disadvantage was distances away from major electric transmission networks. Other challenges included the different types of terrain, from steep mountain slopes to shifting sands, and environmentally protected areas.

To get a better picture of challenges and impediments, data from a variety of sources was incorporated to target specific areas. Detailed direct normal irradiation data was provided by Sempra partner 3 Tier/Solar Reserve. Thirty-meter Shuttle Radar Topography Mission data was downloaded from the US Geological Survey. Environmentally protected areas and world soils were downloaded from the World Database on Protected Areas and Food and Agriculture Organization of the United Nations, respectively. The major data collection challenge was to acquire the transmission network data. Since public or vendor data is nonexistent, transmission lines had to be digitized using aerial imagery with rough general hard-copy maps as guides. Once the impediments were identified and data collected, specific criteria were formed to find the initial target areas.

With an overall map of the country, the areas with the highest potential for solar development were identified and ranked. With these maps in hand, the Sempra development team was able to communicate the potential areas with government administration members of Saudi Arabia and business partners.

Courtesy of Sempra Energy, Solar Reserve.

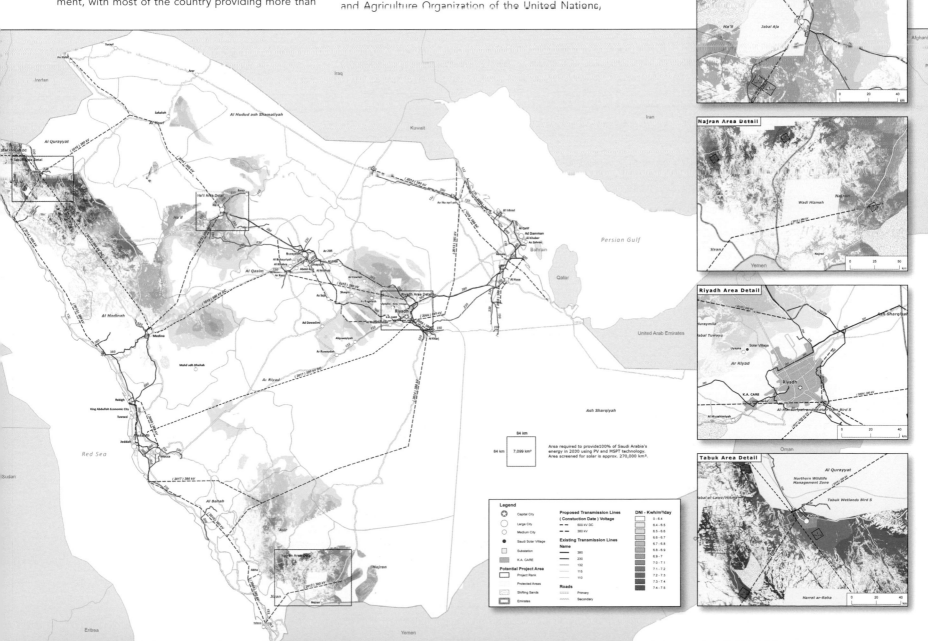

Geo-Schematics Automation of Schematic Layouts for Outage Management and Emergency Response

RMSI
Noida, India
By Amit Rishi

Contact
Amit Rishi
Amit.Rishi@rmsi.com

Software
ArcGIS Desktop 9.2, ArcGIS Schematics

Data Source
RMSI

An integrated solution with GIS and schematics allows the user to retrieve information through selection from schematic to geography and vice versa. Schematics complement GIS for the design, construction, and management of networks, meeting the daily operational management requirements of distribution utilities irrespective of size.

The map shows schematic networks configured on the enterprise GIS database in an ArcSDE environment. Electrical symbols are used to represent various elec-trical and electronic devices in a schematic diagram of an electrical network.

RMSI specializes in geospatial information and data management, with extensive experience in the utility sector. GIS data and application development solutions include creation, conversion, and management of large volumes of spatial and nonspatial facilities and asset data, work order posting, data cleanup, data conflation, and technology migration.

Courtesy of RMSI.

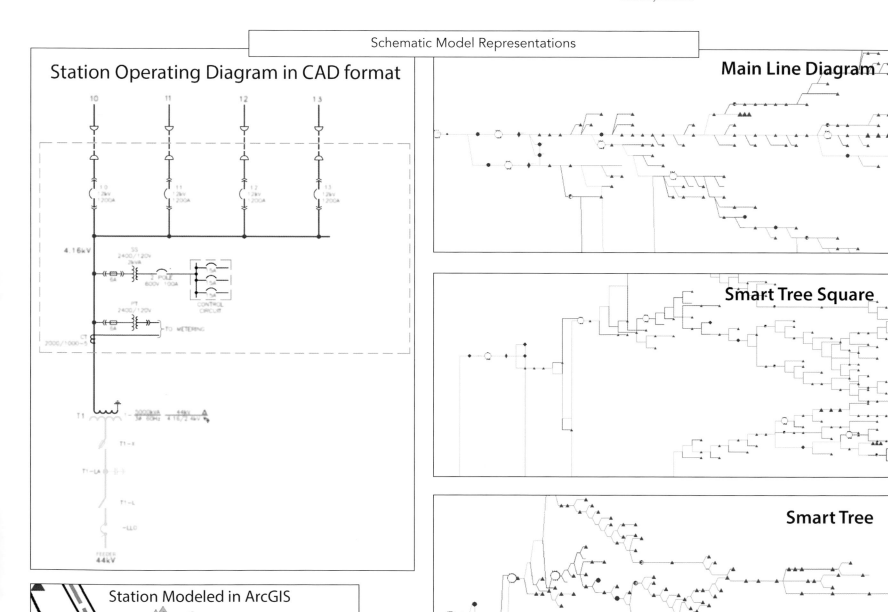

Schematic Model Representations

Station Operating Diagram in CAD format

Main Line Diagram

Smart Tree Square

Smart Tree

Station Modeled in ArcGIS

Maintenance Responsibilities for Drainage Facilities in the Albuquerque Metropolitan Area

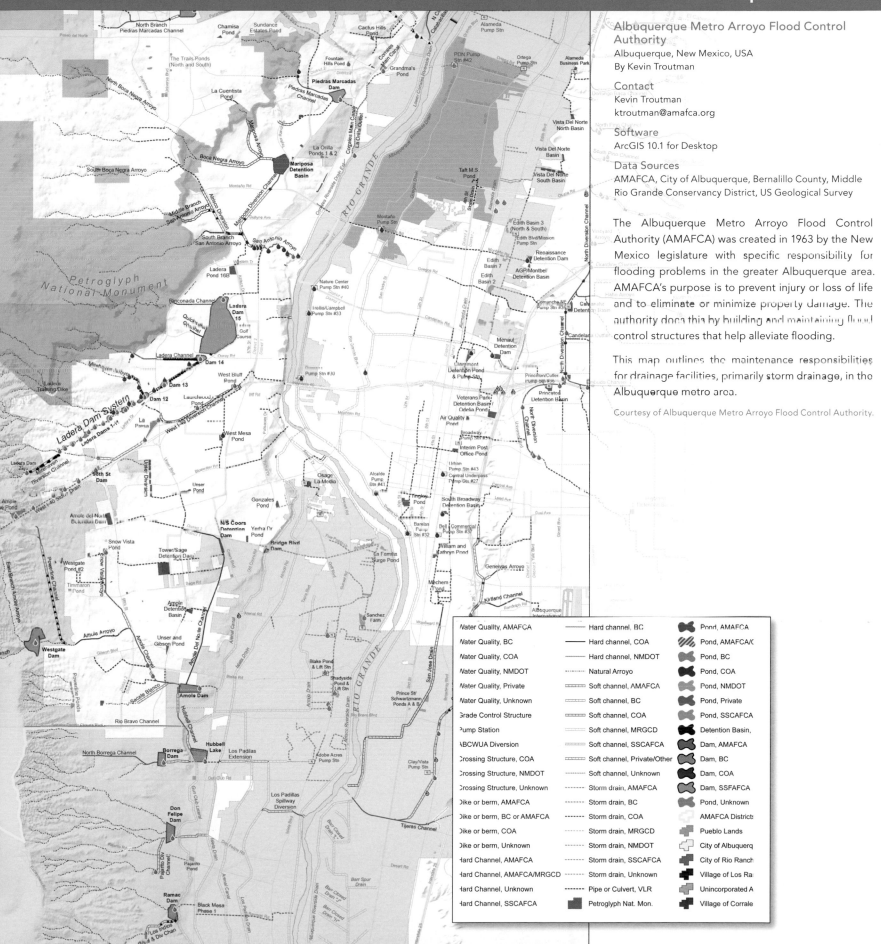

Albuquerque Metro Arroyo Flood Control Authority
Albuquerque, New Mexico, USA
By Kevin Troutman

Contact
Kevin Troutman
ktroutman@amafca.org

Software
ArcGIS 10.1 for Desktop

Data Sources
AMAFCA, City of Albuquerque, Bernalillo County, Middle Rio Grande Conservancy District, US Geological Survey

The Albuquerque Metro Arroyo Flood Control Authority (AMAFCA) was created in 1963 by the New Mexico legislature with specific responsibility for flooding problems in the greater Albuquerque area. AMAFCA's purpose is to prevent injury or loss of life and to eliminate or minimize property damage. The authority does this by building and maintaining flood control structures that help alleviate flooding.

This map outlines the maintenance responsibilities for drainage facilities, primarily storm drainage, in the Albuquerque metro area.

Courtesy of Albuquerque Metro Arroyo Flood Control Authority.

Legend:

Water Quality, AMAFCA
Water Quality, BC
Water Quality, COA
Water Quality, NMDOT
Water Quality, Private
Water Quality, Unknown
Grade Control Structure
Pump Station
ABCWUA Diversion
Crossing Structure, COA
Crossing Structure, NMDOT
Crossing Structure, Unknown
Dike or berm, AMAFCA
Dike or berm, BC or AMAFCA
Dike or berm, COA
Dike or berm, Unknown
Hard Channel, AMAFCA
Hard Channel, AMAFCA/MRGCD
Hard Channel, Unknown
Hard Channel, SSCAFCA

Hard channel, BC
Hard channel, COA
Hard channel, NMDOT
Natural Arroyo
Soft channel, AMAFCA
Soft channel, BC
Soft channel, COA
Soft channel, MRGCD
Soft channel, SSCAFCA
Soft channel, Private/Other
Soft channel, Unknown
Storm drain, AMAFCA
Storm drain, BC
Storm drain, COA
Storm drain, MRGCD
Storm drain, NMDOT
Storm drain, SSCAFCA
Storm drain, Unknown
Pipe or Culvert, VLR
Petroglyph Nat. Mon.

Pond, AMAFCA
Pond, AMAFCA/C
Pond, BC
Pond, COA
Pond, NMDOT
Pond, Private
Pond, SSCAFCA
Detention Basin,
Dam, AMAFCA
Dam, BC
Dam, COA
Dam, SSFAFCA
Pond, Unknown
AMAFCA Districts
Pueblo Lands
City of Albuquerq
City of Rio Ranch
Village of Los Ra
Unincorporated A
Village of Corrale